DX POWER
EFFECTIVE TECHNIQUES
FOR RADIO AMATEURS

DX POWER
EFFECTIVE TECHNIQUES
FOR RADIO AMATEURS

EUGENE B. TILTON, M. D., K5RSG

TAB BOOKS Inc.
Blue Ridge Summit, PA 17214

THE AMERICAN RADIO RELAY LEAGUE, Inc.
Newington, CT 06111

FIRST EDITION
FIRST PRINTING

Library of Congress Cataloging in Publication Data

Tilton, Eugene B.
  DX power: effective techniques for radio amateurs.

  Includes index.
  1. Radio—Amateurs' manuals. 2. Amateur radio
stations. I. Title. II. Title: D.X. power.
TK9956.T553 1985   621.3841'66   84-16436
ISBN 0-8306-1740-X (pbk.)

Contents

The opinions expressed in this book are the author's and do not necessarily represent the views of the American Radio Relay League, Inc.

Acknowledgments

I would like to acknowledge use of material from the following excellent books:

DX IS! THE BEST OF THE WEST COAST DX BULLETIN; original author: Hugh Cassidy, WA6AUD; editor: Charles T. Allen, W5DV; publisher: Charles T. Allen and James M. Allen, W6OGC.

ALL ABOUT CUBICAL QUADS by William I. Orr, W6SAI and Stuart D. Cowan, W2LX; publisher: Radio publications, Inc., Wilton, CT.

My thanks also to William C. Fields III for permission to use a few of the very many funny words of W.C. Fields.

I would like to thank Andy Calandria, K5MVP, for his help in processing and printing many of the photographs in this book.

I want to express my deep thanks to my wife Nell, who stuck with me through the whole project and who typed the manuscript.

Introduction

If you fancy yourself intrigued by the excitement of long-distance worldwide Amateur Radio communication (DXing), this may be the most enjoyable book you will ever read. If you are a newcomer to the ham bands, this book will help you decide a vital question: whether or not to be a DXer. Then, if you decide DXing is for you, it will help save you considerable learning time. If you are already a beginning DXer, it will greatly aid in teaching you skills and save you the hassle of learning things the hard way. If you are in intermediate DXer (not in terms of number of countries but in terms of *experience*), then this book will advance you quicker and give you an edge earlier in competing with seasoned specialists who know how to get things done on the air. Finally, if you are a seasoned DXer, you will really enjoy this book—not for new knowledge but for the insight into yourself which makes all real DXers tick. This is because this book is totally honest about DXers' motives and actions, and the seasoned DXer will chuckle silently reading about the secret truths which comprise the DX state of mind.

Why should you buy this particular book? Why is this book different from other articles or books about DXing? The answer is simple, as just stated. This book is honest. True DXers are aggressive. And highly motivated they have to be. This book admits these facts and describes techniques which are effective and get results. It exposes the secrets all DXers try to hide!

Having said that DXers are aggressive, let me clarify one point.

This is not bad. DX, unlike many facets of Amateur Radio, is competitive. *It is a sport.* The force which drives a DXer is no different from the spirit that compels someone like John McEnroe. McEnroe, I suppose, is a relatively polite and nice individual *off* the court; yet when he is *on* the court, he wants to win, and he wants his opponent to lose (the self-motivated facet of the driving force). He backs this up by playing a game which, if he's hot, will decimate the other chap (the aggressive component of the driving force).

Now, don't draw any wrong conclusions with the use of words like "aggressive." I am not going to advocate illegal maneuvering. To the contrary. Legal operation is much more effective in DX work, and I am a proponent of good DX tactics. However, a pileup is an unfriendly place, and it's unrealistic to assume that a gingerly, delicate approach is going to get you through.

Besides covering operating techniques which are already in use by your competition, this book will guide you in how to realistically set up your station. This book carefully explains the *why* of such essential operating aids and then shows you the *how to*.

I'll try to accomplish this in an efficient and humorous format. For, as it has been said many times, DXing is far too important to take seriously!

The author in his shack.

1

About DX and DXers

There is no aspect of Amateur Radio quite like DXing. Its devotees are the most fanatic, dedicated and stubborn Amateur Radio operators on earth. They are also the most self-possessed in their attitudes, because they consider DXing to be the elite phrase of the hobby. DXers are unlikely to admit this for obvious reasons, but anyone who has accidentally landed on a DX frequency and called a simple QRZ or CQ has learned very quickly that serious business is being conducted there. It is quite obvious to the "stumbler" that the DX operation has priority over everything else. It is important to understand that DXers actually believe that they own or possess a frequency where DX operating is on going.

Hugh Cassidy, WA6AUD, writing in the *West Coast DX Bulletin*, knew this and called DXers "the Deserving." Cass, as he is called, wrote repeatedly on the nature of the DXer, and the following short rhyme expresses his sentiments succinctly:

> *This is the law of DXing*
> *That only the strong shall thrive;*
> *That surely the weak shall perish,*
> *And only the the Deserving survive.*

Now that we have established the gospel that DX embodies "righteousness" and "goodness" and "the best," it is imperative that we define the scope or type of DX activities that will be described in this book. By DX I am referring to high-frequency DX, or HF DX—working Amateur Radio stations in foreign countries. There is also 6-meter, 2-meter, satellite, and FM DXing. These are entities in their own right but are different from HF DXing, which involves the largest group of DXers and which is the subject of this book.

HF DXing takes place on the 160 through 10 meter bands and may possibly take place on the new WARC bands (at 10 MHz, 18 MHz, and 24 MHz) when these frequencies become fully operational in the future. Generally, the HF DXer (hereafter simply referred to as DXer) is pursuing the DX Century Club (DXCC) awards program, sponsored by the American Radio Relay League. For this reason all DXers should belong to the ARRL. The supreme goal of the driven DXer is ultimately to make the (pause, breathe deeply, genuflect if you must) *HONOR ROLL*. This ultimate DX achievement is accomplished by having within ten countries of the current number of "undeleted" (still recognized) countries on the ARRL Countries List. Deleted countries are added to make the grand totals for the general DXCC program. Consequently, old-timers active for years frequently have staggering counts greater than 350 countries. Newcomers blistering through their first 100 then 200 countries are frequently not impressed with these old-timers' figures, thinking they'll have them all within a few years. As they charge above 250 and the fields begin to turn barren, however, even the 300 total takes on the appearance of the mountain it really is. See Chapter 8 for details on the DXCC program.

The result of all the work and sweat and frayed nerves it takes to achieve these totals is to have your call and country count printed in *QST* (the monthly journal of the ARRL) for all DXCC participants. Because they are listed in order of the count, which only the ham himself knows, almost no one else in the world is ever aware that W5 such-and-such has a total of 278 countries. DXers by nature are only able to retain in their memory accurate count of their own totals. True, they know that Joe in their own DX club is the one to beat because he's near 290, but a part of the DXer's brain is constantly activated, which prevents ever recalling another DXer's actual country count. Therefore, thousands of DXers hit *QST* each year, look up their own call and total and then close *QST* with a warm satisfied smile never to see the other thousands of DXers listed with their own totals.

Why does a rational human being do this? Why does someone yell into a microphone literally for hours only to say to someone else, "You're also 5-and-9. Thanks for the new one"? What makes a person sit with headphones that rattle the DXers skull for hours with nerve-tingling code in a CW pileup? Frankly, I don't know. But believe me, it's for real.

AN IRREVERENT LOOK INTO DXERS

Let's examine the DXer for a moment. It is already well known that DXers are the brightest and most agile aficionados of the radio sport. Given this fact, why do they torture themselves so much? Is it that struggle gives rise to strength and pride? The answer to this is yes; yet that is not the only component of a DXer's drive. There is another facet, even more important and also mysterious. You see, the DXer has a disease. A DXer is bitten by a mysterious, invisible organism that results in a disease process known as "DXitis." Far more serious than such minor maladies as appendicitis or meningitis, DXitis is terminal. As a practicing physician and as a member of the medical profession, I can personally attest to the fact that there is no known cure for DXitis. Even super doses of the newer antibiotics, which can knock out many other dreaded diseases, won't even touch DXitis. A brief summary of signs and symptoms will aid those who, for any reason, need to make the diagnosis.

If a DX club repeater frequency being monitored anywhere in the home is activated, the DXer responds with a snap of his head and a remarkable extension of the ear closest to the VHF rig. Reason: A country and frequency may be put out by a club member who is tuning, or some gossip about how so-and-so got his 7Q7 QSL may be aired. A DXer cannot afford to miss this type of information. To see a person's head move so fast and to see a human ear stretch is really a marvelous sight.

If a DXer comes across a pileup where the DX station is not identifying himself, and, in addition, the stations working him are also not identifying the DX station, there is a problem. The DXer does not know who the DX station is. Upon realizing this, the DXer always develops a deep, anxious, tense frown and begins perspiring profusely. He also becomes very grouchy.

If a DXer stumbles onto a spread-out pileup indicating that the DX station is operating split frequency, and the DXer can't find the DX frequency, his hands and body start to vibrate and shake violently, resulting in erratic tuning of the HF knob. The result is that finding the DX station is even more difficult.

If a rare DX station is sending CW faster than a DXer can copy, the DXer develops what is called the "Acute Death Fright Syndrome." This syndrome is very serious and can only be relieved in one of two ways: The DXer suddenly realizes that the DX station is sending slow enough so that at least he can recognize his own call if he gets through or someone announces the DX call on VHF, and the DXer realizes it is a station he doesn't need.

Family and friends of a DXer may see other, more subtle signs when the DXer is not by his rig. On a trip or vacation the DXer's spouse may notice, at meals for example, a silent far away stare and a glaze on the DXer's eyes. This is because the DXer knows an expedition is on and even though he's on vacation he is absolutely miserable. Friends who meet with a DXer regularly—say, for a weekly golf game—often find the DXer late (if he shows up at all) with peculiar excuses such as, "I had something extra to do at the office." (Who has ever heard of a golfer missing a game for work?) The friends of a DXer may take these excuses for years, never realizing there is a deeper, stronger call: the fatal pull of DX.

From the above it can easily be appreciated that DXing is serious business. To a certain extent all amateurs flirt with DX. Virtually every HF ham has experienced the thrill of working something in at least a semicompetitive situation. Yet many hams then shy away from pursuing DX full time. Some are genuinely turned off by the hard-boiled nature of the struggle. Many others, however, are lured to the edge of the fire but are afraid to play with it. This is understandable. Heavy DX work is downright intimidating to the newcomer, and though knowledgeable hams may imply that they can work DX if they want to, detailed knowledge for effective, continued DX progress is not well understood. Yet this does not mean that the desire is not there. The preceding paragraphs were written partly for fun but also to see if you have any of the secret yearnings to engage in DX seriously. If so, take the plunge!

ONSET OF THE INFECTION

DXers become ill over a range of time periods. Some are slowly drawn to it as a mountain fog creeping down the valley. Others are hit suddenly and with stunning force like lightning. Such is my case. After being active for over 20 years I can still recall the instant in which I went from indifference to addiction. At that moment a strange bright light shone over my rig, and an eerie voice came over the speaker pinning the S-meter, "You are stricken!"

It was in 1979 and all kidding aside it happened this way. The combined 3C1 (Equatorial Guinea) and 3C∅ (Annobon) operation was on. 3C1 came on the air first, and I can remember hearing the wild pileup and saying, "Ho-hum; who needs to kill himself over that?" I went sailing instead. Subsequently the group went on to Annobon and put *it* on the air. I can still clearly remember sitting at the rig, fascinated in a new way with the pileup, hearing the smooth way the DX station handled the districts ("I'm listening for 5s only, 210 to 220"), the crisp 5-and-9s he handed out one after another and the sudden consummate drive that I had to work that station. All of a sudden there I was in the middle of an enormous pileup, calling my head off and knowing none of the techniques of tailending, rotating up or down depending on the DX station's pattern, sticking stubbornly to the edges of the frequency spread, picking the "hole" or less-used frequency, etc. And knowing none of this I was indeed fortunate to get one contract, yet without any backup contacts on other bands as should be done on multiband DXpeditions. (How many almost nervous breakdowns have occurred following solitary contracts while the DXer questions the way the DX station said his call, wonders if he was inadvertently not entered into the log or agonizes if they'll lose the one page that he's on? I recommend investing in antacid companies located near some metropolis with a large number of DXers.)

Fortune smiled on me that time. I eventually got the 3C∅ card, which was a nice one because even though it has been on since, it had not been on for approximately seven years when worked. It was at that juncture that, naturally, I pulled out all the DX cards I had collected over 20 years with random QSLing and bureau mailing. I had a big stack and, I was quite certain, I'd have DXCC with ease. I figured I'd just sort them out by country, eliminate the duplicates, send off for the application form, list them, and get that DXCC. What a way to start! It takes some people six months probably to get DXCC starting from scratch.

Then shock! I counted those cards over and over, especially rechecking the Russians because their calls appear so similar, but no matter how I stretched it my grant total after *two decades* of operating came to less than 60 countries. Sixty. I had to work and confirm over 40 additional countries just to apply for basic DXCC. I couldn't believe it. And yet, though disappointment reigned at that moment, it is now clear, retrospectively, that pulling these old cards out and counting them seriously for the first time was an important sign. Just as the light shone over my radio and the voice spoke at the onset of DXitis, here too began a subphase of the disease: the need to have worked countries "in the fist" (confirmed).

Have you ever counted the countries you have confirmed? Now, I don't mean the occasional namby-pamby count after which the exact total is forgotten. I mean the feverish, anxiety ridden, nervous thumbing, blood-pressure-raising DXer's *sacred* count. A count that takes you to a countries list where you note the response: pride perhaps, disappointment maybe, and which is always followed by a pressing urgency to increase said sacred count. (There is a DXing controversy over which is the most fulfilling: the QSO or the QSL. This will be covered later.) If you have ever made the sacred count, then there is a distinct danger that you have or will develop DX tendencies.

In addition, have you ever culled your cards in high anxiety? Are you genuinely irrational when a new one arrives? Do you fondle your cards and smile gleefully, knowing someone else needs something you have? If the answer to these questions is *Yes*, then I pronounce you very close to the state of DXness. I consider this state normal, though a bulk of hams which I designate as "other" may disagree. Finally, the state of DX is more than normal; it is also truly noble.

SOME MISCELLANEOUS FACTS ABOUT DXERS

When you meet a DXer in person or bump into one casually on the air, one thing that is usually noticeable in his calm demeanor and remarkable aplomb. DXers are cool, man. They laugh about the rough pileups and tight skirmishes they have had. They talk about how they worked the 3X barefoot ("Called him one time while waiting for the linear to warm up and got through"). When a DXer talks about the new card that just came in the mail, he assures you that he didn't worry for a moment that it would really come. (There will be a section on the specialized art of how to *receive* QSLs.) When you meet DXers at their club dinners or ham conventions, they radiate confidence with broad smiles and authoritative voices. Truly, DXers represent the upper echelon of discipline, strength, and calmness. Right? Wrong! DXers are calm *after* the fray, not before or during same. True, some new countries are grabbed easily. Many are not, however, and during these times, deep in cubbyhole shacks with nervous faces and eyes unseen by others, much wrath takes place. It is in these private recesses that blood pressures and pulses soar, mental anguish is overwhelming, cursing (hopefully with the VOX off) and foaming at the mouth is rampant. Fear and aggression compete as the DXer watches a DX signal QSB down, then come back up and peak as he now yells for every watt the linear is worth.

And when the contact is made? "Aww, it was easy. Nothin' to it." And yet there is sweat on the back and arms of the chair, perspiration spots on the log, fingernail marks on the desk, and blood on the microphone. If a DXer ever denies that he is emotional in a pileup, ask him how he felt after a long struggle over one he *missed*. Although he won't admit it, you will be able to see him wince for a moment. In his eyes you will see that anguished feeling over such a loss. For when a good one passes by, there is weeping and gnashing of teeth.

Another miscellaneous characteristic of DXers is a remarkable sense of deep humility. It is remarkable because it is false. Humility has two angles in a DXer. One is expressed gratitude for making a difficult contact or getting a rare QSL. The other is humble mumble about one's country total. ("Shucks, I only have 283.") Neither of these expressed thanksgivings is taken seriously by DXers. They are for public consumption only. While in a technical sense a DXer is grateful for his achievements, he is not for one moment humble about it. A real DXer believes he *deserves* contacts and QSLs. It is his *right*. And when one is right, one needs not be humble. It was no accident that when Cass wrote about "the Deserving" in the *West Coast DX Bulletin*, he always spelled it with a capital D.

PERSPECTIVES

Is serious DXing for you? Many can answer such a question with a simple yes or no. Others would have to answer with varying degrees of uncertainty. As previously stated, most hams have flirted with DXing and there are many casual DXers. Serious DXing on the other hand is a long-term commitment. It is also very time-consuming. In deciding whether or not to be a serious DXer such facts must be completely understood.

If one is to compete as a serious DXer, one has to realize that the struggle is not over after the arduous tracking and snagging of a particular station. There are many, many more that will require considerable effort. Serious DXers don't rest on their laurels or duffs. They get back to the game fast for if one does not, an opportunity may come and go that won't reappear for years.

Yet, one should not resist DXing because of time factors. Don't avoid this exciting phase of the hobby because you are legitimately busy. There will be, of course, natural envy of those who have more time than you have. You might miss a few because, in the end, professional and other considerations will and must supersede. In the final analysis, competition in a hobby is relative to factors that are very individual. If you are a neurosurgeon or an electrician logging more overtime than you're logging on the bands, you will work fewer stations than the retiree with no time limit. If you have a straight jacket job with no flexibility, you will anguish over your self-employed friend who can extend his lunch hour to make the D68 list. But, within your limitations, your percentage of countries worked that are available will still reflect your own skill and effort. And finally, with perseverance you won't be far behind those with unlimited time. With a few exceptions, just about every DX opportunity can be mastered—if you try!

Trying is a good word here because even if you try very hard, DXing is still very trying. Frustration and strained nerves are routine, particularly as your DXCC count rises and difficult, time-limited DX operations are fiercely contested. It is here, at this mental juncture, at this crossroad of decision—whether to jump into this psychological cataclysm or not—that DXers and outsiders meet and test each other with their own points of view. DXers are content that they are participating in a sacred contest. Outsiders, whether they admit it or not, are drawn to this spectacle and do one of three things: 1.) Join in part time; 2.) Join in full time; 3.) Don't join in at all. Regardless of their reasoning, however, all approach DXing and consider it. This is not true of other phases of the hobby. (Consider the small percentage of hams that has ever investigated—much less operated—satellite, slow-scan TV, 6 meters, RTTY, etc.) I don't think anyone will argue that DXing cannot claim an enormous number of active participants—and an even larger number of secretive, would-be participants (closet DXers).

Thus, DXing is a vital and very important aspect of Amateur Radio deserving of your attention. And it is the purpose of this book not only to show you how to DX effectively but to help you make the psychological decision to enter DXing in a serious, long-term way. As stated previously, many may scoff and say such a decision is not so complicated. They mention the pileup they were in last week and explain that they could be real DXers if they want to be. Some even think they are real DXers after a successful battle in a pileup. But one pileup—or two, three or forty—does not a DXer make. Very specific operating skills are needed for the long haul. In addition, a special determination is necessary for the mental edge that keeps a DXer close to working everything that has come on since his interest began. This is not easy. Ask those above 250 countries.

Well, certainly there is grist for the mill here. Beginners and seasoned DXers alike should be warming up to the special thrill that is DXing. For it is written in the somewhat risque words of Cass:

> *Not for thee the joys of family,*
> *or the siren call of sex.*
> *Your vocation, you've decided*
> *is to chase the rare DX...*

We shall see.

2

In The Beginning

SCENE: New York, the production studios of one of television's most popular programs of all time, a Sunday night newsmagazine show:

Voice on intercom: "Mr. Rooney, 30 seconds to tape."
"Right."
From the director, "Andy, turn to your left a little."
"This all right, Maurie?"
"Fine, Andy. Roll tape! Five seconds, four seconds, three seconds, two, one . . ."
"Hello. You know what bothers me? DXers. Yeah, those hams that you hear so much about. You can't even understand what they're doing. I heard one fellow saying he had contacted YU, DU, VU, and TU but still needed XU. His friend said he had G, F, and W but had not contacted I. What are these guys talking about? The alphabet? How can you contact the alphabet? And I thought Californians were weird.

"Then one day I heard one of them say that he had 248 in the fist. What does this mean? Is he physically grasping the number two hundred forty-eight? I looked in his hand and I didn't see anything.

"And regular hams at least say 'hello' when they speak to others on the radio. DXers only say something like 'five-nine.' What is this? A subversive code? Is this the signal to take over Poughkeepsie? Is Washington, DC next?

"But the strangeness doesn't stop here. I heard a DXer once say, 'You're five and three, five and three, that's fifty-three, QSL?' My goodness. Five and three is *eight* not fifty-three. Can't they add? These fellows must be deficient in math.

"Yes, DXers are strange. They're like those toothpaste tubes that don't roll up right, a constant annoyance. I went into a DXers shack once. *Shack* is a good word here for even though his house was nice this room was utter chaos. Wires and equipment and stacks of papers and cards were everywhere. I wonder what the divorce rate of DXers is.

"Then another DXer wanted to show me his finals. I thought he meant exams or something. He straightened me out quick. He meant tubes. Well, I thought tubes were not bigger than beer bottles but he showed me something that looked like a couple of five-gallon water jugs. When he turned it on, all the room sort of glowed. I thought we were going to evaporate. But this DXer just stood there calmly with a strange smile and a peculiar gleam in his eyes. He wasn't worried—he was happy. He was possessed.

"There are other things that bother me about DXers. They remind me of paper clips that bend too easily or postage stamps that don't stick—you know, the kind of things that really get under your skin. For example, I saw a DXer get a 'sorry, not in log' QSL response once. He got very upset and bit his dog. Another got squeezed out of the BY pileup. he sacrificed his daughter to 'appease the gods.' Goodness, these fellows are crazy! I'd hate to see what would happen if one of them lost all his QSL cards collected over the years. Well, that's it. That's enough of DX and DXers. Good night."

Brief silence.

Then, the director: "OK. That's a wrap. Put it away guys. Thanks, Andy."

"OK, Maurie. Good night, fellas. Oh, Maurie. Close the door behind you."

"Right, Andy."

The crew's footsteps shuffle away. It is quiet in Andy's office. A few buttons are silently switched on. A microphone is grasped. A crusty, familiar voice calls, "CQ DX, CQ DX, this is Whisky 2 . . . ."

* * * * * * * * * * * *

Well, I don't think our TV friend is a radio amateur, and perhaps even his bluff naivete is exaggerated. DXing and DX related activities may be foreign to many hams but it's not that strange to most. However, true-blue DXing is different, and depending on the level of experience, DX ways and means may appear complex and confusing.

When amateurs first become active, everything seems like DX. To the beginning ham in New York, a contact with California is definitely considerably distant and considerably exciting. When one contacts his first European, the thrill of having spanned the ocean is palpable. Soon, however, such contacts become commonplace and enthusiasm levels off. Calls like HA and PY, once intriguing, are now routine. Then, such a ham comes across a couple of DXers discussing countries. He may suddenly be astonished to hear calls that require checking an atlas to find out what they are. So just when one thought he had the prefixes of most countries that he usually hears on the air in his memory bank, he suddenly hears DXers talking about S9, A4, or 7Q. Fearful, he looks these up so he's not left in the cold world of the unknowing. There is some consolation because at least he recognizes the countries (Sao Tome, Oman and Malawi). But later, when someone else is discussing 3B8, D6, and C2, he may be surprised to learn that there are actually countries called Mauritius, Comoros, and Nauru.

Other elements of the DXer's world frequently seem mysterious to the beginning amateur. He may hear a big pileup with the DX station competing with the very stations calling him. In this mayhem people are nevertheless getting their reports a high percentage of the time. Then, our new ham may tune in a list operation where everyone is working an extremely weak station. Each time the DX station transmits, there is only a faint mumble, a mere "wind-whisper" of a signal. But most of the DXers are accurately repeating their reports without any help. How do they do it?

In a casual QSO, a DXer may relate the complicated way he got a card by purchasing that country's stamps here in the U.S. and then deliberately fixing them to a self-addressed, stamped envelope to help prevent theft by postal officials in that country. Or one may hear the tip to avoid writing call signs on the envelopes; call signs tempt postal personnel in certain countries to open the letter and remove any money or International Reply Coupons (IRCs). Further, our new ham may hear the instructions for the special way to address the envelope for a certain card so that it has the best chance of not being confiscated and destroyed by that government's officials. And there's the knowledge of when one should never enclose U.S. money when QSLing, because receipt of foreign currency is illegal in certain countries. The ham involved could get in actual legal trouble with his government if discovered. All of this certainly does not represent average QSLing technique. How does a ham get though this intricate maze to get the cards from these hard-to-get lands? Still other aspects of DXing may befuddle the newcomer. DX bulletins and DXers alike often state that a certain country or DXpedition will be on the "usual frequencies." To the uninitiated this can be a perplexing statement. Or how about when there's a hunt for a hard-to-get station, and you hear several DXers say, "Let's QSY to where he usually hangs out." Where on earth could this be?

DXers tend to guard their knowledge jealously and often speak cryptically to their friends to prevent wide dissemination of valuable information. It's not unusual to hear something like this: "Joe told me that the license came through. It looks like they'll be leaving on the date I told you previously. When they start up, they'll be on the special frequency." There may be a response such as, "OK on that. If so, I better notify Jack on the West Coast. He's in touch with Peter in the Philippines who can call them first on the other frequency." Can you imagine a neophyte DXer bumping into this? Not only can he not glean a time or frequency from this exchange, but even if he listens a long time, he will never hear just what *country* is being talked about. People will break in and ask mysterious questions about the boat being used, types of antennas, and so on, but no one ever mentions the country. They all seem to "know." Our neophyte finally gets the courage to break in and ask just *what* country is involved. But he is told, sternly that at this stage everything is tentative and the information cannot be given out at this time, less confusing rumors start to circulate. The neophyte, always grateful to talk to big time DXers, replies that he understands and then thanks those on frequency who had given him nothing.

Can all these mysterious activities be unraveled? The answer is simple: yes. Experience is one way. This book is another. I shall endeavor to share the wealth in these pages. There will be considerable information, for I am not just relating my knowledge but also the knowledge of other more experienced DXers.

YOU HAVE TO START SOMEWHERE

Even if you don't want to be a DXer, even if you don't want DX QSL cards, and even if you want to avoid all traces of DX activity, you can't. For when one has that license and gets on the air, he is automatically headed for about 50 countries worked and maybe 30 or thereabouts confirmed. How is this so?

By simply calling CQ, a certain number of common countries are going to answer, and you'll eventually run up a small DX total. And even if you swear you will not QSL, a percentage of QSL cards will float in—some direct and some through the ARRL bureau in those envelopes you had sent in as soon as your first license arrived in your mailbox.

Most hams don't avoid DX or course, but they start out casually about it. They happily exchange QSL cards. There is usually a pile of US cards and a separate DX grouping. The number of foreign QSLs may appear to be high, but all those duplicate German, English, Italian, and Latin American cards make the DX pile only *look* tall. The actual country count is usually fairly small.

Somewhere along the line, the routine DL and I contacts become a little blase. Pileups, which were at first carefully avoided, start to look a little interesting. Fore example, there's an 8P down in Barbados or a UB in the Ukraine that has got a little action going. It's not too heavy. After listening a while, the tenderfoot notices most of the callers are eventually getting through. Although nervous, all beginners want to test the water and usually will try a small pileup. With any luck at all, they get through. Immense pride wells up inside. There's usually another pileup after the first one. For some, addiction is starting. For others, there is mild temptation competing with a kind of fear and contempt over fighting for a QSO. It just does not seem honorable.

The previous experiences are common with just about all hams operating HF, for DX is simply there. After the flirtation, however, the paths vary. Some go immediately into full time DXing, some avoid serious DXing forever, and some pass through other phases but wind up back in DXing later—much later sometimes.

One thing is certain, however. After simple pileups, two of the three groups just mentioned move on and up. And when a new DXer—whether a long time ham or not—participates in his first really serious pileup, one thing happens: he is very impressed by it all. A heavy pileup, particularly when one cares about getting the contact, is an emotional experience. The ferocity, the intensity, and the immensity of serious competition for a rare one causes an adrenalin rush and a combination of fear and excitement that is hard to describe. Even experienced DXers, hardened by frenetic combat, still get amazed at the enormity of energy in some pileup situations. The action is fast and demanding, and the exhaustion after it's over is far out of proportion to the amount of physical energy expended.

But it's a long way to the confident level of pileup handling. So how does one get started? What are some tips for beginning? What's it like to chase an OX or a TA when one is tired of VKs and Gs (except for friendly, ragchew QSOs)? What does one feel when one makes his first embarrassing bumble while desperately trying to get his first 9Y? Let's investigate some of these questions.

MISTAKES

As everyone knows, I have been a scholar of proper procedures.
W.C. Fields

As everyone knows, DXers have not. Fields wasn't serious, of course, but DXers *believe* with deep sincerity that they are orderly. They think they follow the rules, written and unwritten, and are astounded if anyone actually criticizes their operating habits. However, tape recordings are easily made and as they say in the World War II movies, "Vee hav you in oower fiiiles!" DXers will tremble with that statement. Rest in peace, my brothers and sisters, I—unlike President Nixon—have destroyed the tapes. Yet, examples of flubs abound. I present some representative situations.

When a DXer is in an intense pileup and wants a contact very badly, he is likely to think he has heard his call sign from the DX station when in fact he has not. And, after boldly coming back and giving the DX station a report, he is likely to hear the following from the loudest station around: "Hey, K5RSG, ya jerk, he came back to K9SG. Now shaddup!" Even though no one can see the bumbler's face, it is in fact very red and that special expression for such occasions is acutely manifest. At this point a healthy pause is usually allowed to permit all those who heard this embarrassing episode to work the DX station and leave the frequency. It should be noted that the bluff call—an attempt to bulldoze one's way into a QSO—is a separate entity from the legitimate mistake. The bluff call is a deliberate technique from which an operator derives only a small amount of embarrassment. That small amount is only due if he gets caught and doesn't get through.

Another great boo-boo is to suddenly find a pileup, hear the call quickly—say ST2AB—pop your call in and get a response. Verifying the DX station clearly (Sugar Tango Two Alpha Bravo), giving the report (five by nine, of course) and closing out with, "so nice to talk to you from the Sudan," one is suddenly surprised to hear the correction, "sorry, old man, that's S *P* Sugar *Papa* 2 . . ." Fortunately the giggles on frequency don't last too long.

One of the worst mistakes frequently goes by unknown at first, not only to the DXer but the DX station as well. The following, unfortunately, is not all that rare. Two DX stations are close to each other. Both are listening up. Stations are calling each, all intermingled in one big pileup, all spilling over onto both DX stations. This is no problem when the DXer has the two DX stations isolated. However, when an inexperienced CW DXer hears a station rattling off CW reports, he can get confused. To add to the problem, more often than not the DX station identifies infrequently and fast. Thinking he has the one he wants in tune, the novitiate (and the experienced DXer as well) calls and finally gets through. But to the wrong one. Two-meter and DX bulletin information confirm that the "right" (unfortunately *wrong*) station was on at the time and frequency, and QSL data is passed along in good faith. I do not know how many times operators catch this mistake and go on to make the right contact, but the number of occurrences of working the wrong station is higher than it should be.

Another mistake is all to frequently not one at all, but when it really is a legitimate error, it too is embarrassing. DX stations frequently work the masses by some type of designated areas such as U.S. call districts. Inexperienced operators who land on frequency and hear the DX station calling QRZ may pop their call in without realizing what is going on. When this is a legitimate mistake, it's usually red-face time again, as there is just no end to the people who will point out the bulb. ("Hey, imbecile, he's on 2s, not 4s! Did ya leave your brains in the sink?") Ahh, the camaraderie of DXing. As stated before, however, calling out of turn is often deliberate. Such action can be risky. Sometimes the DX station scratches violators from his log, not only for that QSO but for others on that DXpedition as well.

As you see, a *faux pas* can invade many an illustrious ham career. Some mistakes can't be avoided—we all have to learn somehow—and others are mortal sins. The newcomer should be prepared for embarrassing moments. The old-timer should set a limit. No matter how experienced the operator, enthusiasm will generate genuine mistakes, and the novitiate should not beg off from the game out of fear. His mistakes will be small and less turbulent. The newcomer, with his inherent sincerity, is usually forgiven. For one thing, the experienced DXer always knows when one of his own kind—under the guise of a mistake—has deliberately fluffed. And if such a maneuver costs someone a contact, there is no forgiveness. Yet, in the end, mistakes happen. They usually make good "sea stories" to be relived at DX meetings and other eyeball QSOs.

ZERO HOUR

Though the starting point on a serious commitment to DXing is often hard to find, there is a line that, when crossed, separates the casual operator from the serious DXer. Sometimes this line is crossed clearly, with acute vision and awareness; at other times it is traversed slowly, as a sort of gray zone or hazy transition, and one suddenly arrives at the other side surprised but nevertheless committed. Because beginnings vary, a so-called *zero hour* is hard to define. For the sake of simplicity, I shall not deal here with when one gets his license. As stated previously, any ham, DXer or not, is going to fall into about 50 countries. So I shall start beyond the 50, after the simpler ones.

Before we can eliminate the easy ones, we should briefly point out what they are. Succinctly, Western Europe, Central and South American, the Caribbean, and the East (Australia, New Zealand, and Japan) provide a core of countries that you almost cannot avoid. From this general family, which contains many different ones too, comes what might be called the "free lunch." After this, the menu becomes just a bit more expensive.

We will assume that DA and EA, UK and VK, YV and SV, DL and ZL, and others of similar ease are already in hand. Thus, zero hour begins beyond these. First look at the DXCC list to see what to attack, keeping in mind speed and efficiency. Take a little world tour to get a general feel of the countries. Then outline a specific strategy for getting started.

Europe

Because all the major countries from Finland to Greece have been worked and confirmed, be on the lookout for some "lesser knowns" that show up *routinely,* though on an intermittent basis. GU (Guernsy), GJ (Jersey), GD (Isle of Man), SV (Crete, not Greece), HB0 (Liechtenstein), HV (Vatican City), FC (Corsica), and 3A (Monaco) are such good examples. Hams actually *live* in these countries. Though DXpeditions sometimes activate these lands, these countries can be found by simply tuning around. This information may not be trumpeted in DX bulletins (except as contest information) or on DX repeaters, so one simply must roll up his sleeves and find them. It's easy and fun, but you must be patient. Another thing to look out for, particularly in contests, is activation of such places as OH0 (Aland Island), OJ0 (Market Reef), or SV (Mount Athos). These European goodies are normally *not* on the air except by DXpeditions and you should watch—through bulletins and buddies—for when and where they are scheduled to appear. Within six months (a year at the most), *all* of Europe should be within reach with, at the most, only an exception or two. For example, although 1A0 KM (Sovereign Military Order of Malta) makes several appearances per year, the demand that builds up in between makes for a fairly nice pileup. A beginner might have a hard time cracking through.

The Americas

A wealth of countries are between Alaska and the waters below Chile. The Caribbean is in this group. From the standpoint of propagation, these are relatively easy catches for hams in the U.S. This is literally our backyard, and the number of countries available here is significant. Depending on where one wants to draw borders around this whole grouping, there are over *60* countries. Over 80 percent of them are readily accessible for QSOs. Some are only rarely activated by DXpeditions. Because of their location, though, even these are usually not hard to get. One should literally milk the Americas for all its worth.

Asia and the Pacific

This is a real hodgepodge of very easy and very difficult ones. The range here is incredible, from ragchewing countries such as Australia and New Zealand, to the heart-stopping thrillers of S2 (Bangladesh), A5 (Bhutan), and BV (Taiwan). The distance is considerable so the pileups for the tough ones are rough. And *many* of the countries in this area are on infrequently (Central Kiribati, Kingman Reef, Spratly Island, and Mellish Reef, for example), and some are not on at all (Vietnam and Laos). Yet the beginner can have a field day here, working Japan, Hawaii, New Zealand, Australia, the Philippines, and the like and occasionally catching a goodie like Pitcairn and the Solomon Islands in between them.

The Middle East

Another good mix presents itself here with simple ones (Israel and Jordan), moderates (Turkey and Saudi Arabia) and "toughies" (Iraq and Pakistan). It has its share of inoperatives, such as 4W (Yemen) and 70 (Yemen Democratic Republic). With persistence, however, you can increase your total here, for even though pileups for countries such as Lebanon and Oman are moderately rough, they are on often enough so that you are usually rewarded on some frequency at some time.

Africa

Work every African you can whenever you hear it, for tomorrow it may not be there—at least the government that permits ham radio operation may not be there. It is difficult to characterize Africa from a ham radio standpoint, because it really is a place of great flux. There are a few easy-to-get countries, such as South Africa, but the majority of nations are unfortunately precariously situated. Governments come and go. So do ham licenses. We used to ragchew with 5A (Libya) and C9 (Mozambique) in another time and friendlier atmosphere. Now, as we know too well, these and many other African nations just don't have ready access by ham radio.

Russia

The Union of Soviet Socialist Republics consists of 18 separate DXCC entities from the standpoint of Amateur Radio. These 18 DXCC countries form a gradual gray scale, ranging from very easy to contact to very difficult to contact. European Russia is a cinch, of course, and an example of the most difficult ones might be Franz Josef Land or Kirghiz. One of the biggest problems in dealing with the Russians is not nuclear missiles, armed forces in Europe, or even QSOs with them, but *getting the QSL card.* It takes years and years to confirm all 18 of the Russians. It also takes tricks, cajolery, cunning, and luck. We hams could teach our State Department a thing or two about patient negotiating.

As far as working them is concerned, place a separate list of the Russian prefixes at your operating position. The prefix determines which country it is, and it is difficult to remember all the little variations of the prefixes. They should be worked as many times as possible because of the already mentioned QSL problem. While one contact with a rare UM might bring a card, several contacts with another will not. Predicting which one is going to be the problem is impossible. Every DXer, when he finally gets all of his Russians in, breathes a long sigh of relief.

Others

A few countries are not included in any of the previous areas. Antarctica, Heard Island, Jan Mayen, Iceland, Greenland, and the Faroe Islands are some of these. Antarctica depends on radio for communication so some ham from some country is almost always there. Some of the other areas in this group are uninhabited, and, in addition, are very remote, making activation difficult and expensive. The cost of activating Heard Island, for example, ran about $40,000 last time. If anyone doubts that DXing is serious business, this should squash that notion.

THE CASE FOR LISTENING

Talking is a cinch. Listening does not come easy to hams (which is why we are probably called such). Let us then explore some reasons to help motivate us into turners or listeners. First, countries such as VU (India), EA8 (Canary Islands), and ZS3 (Southwest Africa) do not draw unusually large crowds and do not stand out on the air. It's easy to find the Glorioso or Juan Fernandez DXpedition because the pileups are so huge. One must tune more carefully, however, to find a humble EL (Liberia) or CT2 (Azores).

Another reason that basic tuning is necessary is because, even assuming one is a subscriber to a DX bulletin or two, these countries may not be even "advertised." No one writes that Montserrat is on or that Fiji is coming up. These countries are, for DX purposes, not sought-after enough. But they are also not on the air like the Englands and the Italys of the world. They require some effort and careful searching.

Another source for *no* information on these countries is the DX repeater. Whether in a club or not, most hams are near enough to a DX club repeater to her the poop being put out. Well, you can be sure that *no* DXer is going to say that the Bahamas are now on 14.195 and listening up. It's beneath one's dignity to put out such lowly information for, as we all know, *everyone* had this one long ago.

So getting started requires working alone. There will be very little aid and assistance, and it's a good idea to simply get going. There are a few benefits in this, however. For one thing, working an India or a Lebanon requires a little bit of knowledge about pileup busting. The hands-on experience gained here will be valuable later on. Another thing is the ability to work anonymously and quietly build up your total. No one goes around bragging that he just worked the Netherlands Antilles. He would be laughed out of town. So you can work quietly. *Suddenly* all those little countries total 100, and you can go to your friends and say you just got your DXCC. Nobody's going to give you a medal for this, but all DXers remember when they got theirs. It's a real milestone that shows some commitment. You will be respected for it.

One final plus in tuning and working alone at the beginning level is that your first on-the-air mistakes are not made before large audiences. It therefore gives you a chance to polish up your act in relative obscurity.

AND NOW A WORD ABOUT TIME

We will shortly be discussing specific DX techniques, but one thing that should be covered first is how to balance your time. If you are ready to use a valuable weekend for DXing, is it more profitable to work 10 easy ones or should you chase the Cocos-Keeling DXpedition that might eat up the whole weekend *and* might produce no QSO? Worse, should you try to make the D6 list (Comoros) on 20 meters or the 5R list (Malagasy) that is simultaneously on 15 meters? These events really happen and must be handled. Here are some guidelines.

In general, you should always try to work a rare country that is activated by a DXpedition and probably will not be on again for some time. While it is surprising that seemingly impossible countries are regularly put on, it is also true that some are not. Given political reality or location problems or both, a rare country could enter into an era of no activity for a very long period of time. It took forever for China to finally come back on. It wouldn't surprise anyone if Vietnam and Laos take decades to be activated again. Thus, when really rare areas like Spratly, Heard Island, or Clipperton come on, go for it at the expense of everything else.

Some countries, such as Tromelin and Rodriquez, are put on habitually. They are not on frequently, however. As a result, the pileups are formidable. If you are inexperienced and not likely to get through, there may be a case for *not* sacrificing a block of valuable operating time. In this situation, call an experienced DXer to see if this is one of those countries that reliably comes on again and again. You can't work everything at once. Because use of time must be selective, it may be occasionally wise to avoid a very competitive situation until you are reasonably certain of producing results.

The final conflict—that of equally difficult countries being on at the same time—is perhaps the hardest choice to make. What do you do when T2 (Tuvalu) and J2 (Djibouti) are on at the same time on different bands? Don't say it doesn't happen. Not only has the equivalent occurred several times to me, but once I encountered the same DXpedition on two bands, each coming toward to my fifth district in opposite directions. One was on 20 meters working W1s, then W2s, and so on. The other was on 15 meters working W0s, W9s, then W8s, and so on. I couldn't believe that this ridiculous situation was happening, that each station of the same country that I really needed would land on the 5s at the same time, thus cutting my chances for a contact in half. But it did occur.

I had to size up propagation quickly and choose my best shot. The country was Kingman. I did get through, but it was mostly luck. I still can't say for sure if the Louisiana 5s, at that moment, were favored on the band that I got through on, or whether I was in the right "hole" at the right time on the wrong band. It doesn't matter now, of course, but the conclusion I'm trying to draw is this: When conflicts are simultaneous, when choosing logically seems impossible, stop thinking. Point a finger in one direction and head there. In the end, definitive action is better than scattered, diffuse action. If you come up wrong and miss the country, remember that it's only a hobby.

Remember that the challenge comes from only countries that are not easy. It is irrelevant if, for example, FG (St. Martin), 5W (Western Samoa) and 5B (Cyprus) are all on simultaneously. These are relatively easy, even if not common. No, the strained nerves come when South Shetland or Sierra Leone or Botswana are on at once and, unfortunately, for only a short period of time. Here is where the true DXer can't help but be anxious and, if sufficiently rattled, several valuable contacts can be missed. Once you have been through this a few times, however, the quick reacting skills that are necessary become developed and the job gets easier. There is also a feeling of considerable pride when you are switching around between several scattered, tough pileups and then get through to some good ones.

BEGINNING TECHNIQUES

Ahh, if we all could be James Bond, slipping into the casino with a luscious lady on one arm, holding a winning hand at the baccarat table, pulling away a million pounds Sterling from arch foe Goldfinger. Savoring the victory, we straighten the tuxedo jacket, swish down the Dom Perignon (1966), and drive off in the Ferrari to the Alphine retreat, finally seducing the beautiful Russian spy while working a ZA on the 100-watt wrist transceiver. That's how it's done, folks. No flaws, no errors.

Well, if it's more like Laurel and Hardy, if one hits himself in the face after grabbing the mike for a rare one, if one sends four dahs and three dits when nervously trying to send the 8 in his call, if one says he QSLs the 5-and-2 on the list operation when the DX station gave a 5-and-4, if one thanks the hard-to-hear DX station for the new one while the same DX station is acknowledging someone else, or if one thinks he's working a rare KC6 in the Western Caroline Islands only to find that the station is a KC6 in Fresno, California, then one has encountered DXing!

Old-time DXers like to think that their awkward days are over. The facts prove otherwise. In the heat of competition, misjudgments are often made. Please note that I'm differentiating operating mistakes—real or deliberate—from the other internal personal goofs that affect no one else but the DXer himself. Hence, I am not talking about calling out of turn but rather doing something else such as listening to the wrong station. There is a distinct difference between an *intrusion,* intended or not, and an *illusion* where one innocently misinterprets or misjudges a situation.

This brings us to the interesting point, which no experienced DXer likes to admit, that even a "professional" makes many of the misjudgments that he made in the beginning. To the neophyte, old-pro Joe seems to be flawless when routinely on the air, especially when recounting ancient battles which were fought with honor and good judgment. Then Clipperton comes on, which old-pro Joe needs. The neophyte hears Joe in the pileups for the first time. And there he is calling when the DX station is transmitting, there he is tail-ending the wrong station, and there he is asking someone on the repeater if he got through. Wow! What disillusionment. Joe's *not* God. Why, he's *human.*

This is a funny thing about DXing. No matter how good one is, no matter the experience one has, when we're in there for a new one the same confusing elements (propagation, QRM, QRN, and the multitude calling) all contribute to mishaps. Another thing to be noticed is that the once cool old timer, yelling for something *he* needs, has an urgency and quiver in his voice that sounds like the excited newcomer. How about that! We must be all the same. And so we are.

The newcomer should shed all fears about DX techniques and realize that, even though he will produce a few more inaccuracies, the experienced DXer is not too far ahead. The difference is that the old pro participates in far fewer competitive situations and thus has less "exposure." I certainly don't mean to imply here that a beginner is equal to an experienced DXer. The old pro knows tricks that will devastate the neophyte. On the other hand, however, the newcomer should not be intimidated into thinking that he is out of his league and can't participate in the game. Different phases of on-the-air techniques are handled as you grow. Enjoy it at each level.

Tuning

We've already said that DXers listen more than they transmit. The newcomer needs to learn to discipline himself and turn away from the urge to babble and turn to the art of hearing. This is one time when it is indeed better to receive than to give.

Tuning is basically not difficult and really only requires the patience to keep doing it when things seem to be quiet or nonproductive. For the beginner it provides many rewards, because there is so much out there. You must be careful, however, because as you pass through different steps, new ones seem to level off. It is easy to yield to the temptation that "all that is reasonable" has been worked.

The first such step comes when you have 150 to 175 countries. There suddenly doesn't seem to be anything new out there. This is far from true, of course. Persistence in tuning and watching for DXpeditions are all that is needed to keep going. The next major blockage appears around 250 countries when the stream really seems to run dry. It is true that things actually do toughen up at this point, but here are some surprising facts.

The current DXCC list recognizes 316 countries. Approximately 80 countries are not routinely active. This leaves *236* that are. Although some are not easy, they are there, they are active, and they are accessible. Of the 80 that are not on the air, slightly over one-half, 40 plus, are put on *routinely*. Examples of this group include the frequently activated Aland Island (OH0) and Sable Island (CY9). Some like Kingman Reef (KH5K) are put on less frequently but do make it every few years. All this adds up to about

275 countries that can be considered readily available over a period of several years.

There are just over 20 others that are activated only infrequently. This would include such places as Spratly (1S) and Heard (VK∅). And finally, 11 countries that are currently totally inactive. One example in this group would be XW, Laos. Major changes will have to take place before some of these lands will accept Amateur Radio operation.

All in all, however, there is a lot out there. Diligent tuning will bring it to you. At first, it seems strange when you give up routine transmitting. Yet, in time listening becomes natural. You hear all sorts of interesting things on the air that went unnoticed earlier. As you bump into DXers talking about rare countries or into unusual QSOs or even non-ham transmissions, listening becomes enlightening and entertaining.

How to Listen

Twisting the VFO knob is easy, of course; however, for the beginner, simply running up and down the band is not the same as precision tuning. First, listen *carefully,* particularly when stations are weak. Stop for a minute and find out what station or stations are on, even if you have to strain and concentrate. This seems so logical and obvious that the advice almost sounds foolish. Remember that beginners are usually used to rapid tuning (they look for the loudest CQ for a comfortable ragchew). Careful, patient tuning does not come naturally.

Listen as often as possible to foreign stations in order to develop a feel for different accents. Then, when you find a weak, needed country, you have a better chance of identifying the station and getting the proper report and QSL information. Though sometimes the stations working the DX station identify the call for you, frequently they do not. It's not unusual to hear round after round of DXers working a DX station by only identifying their own calls. It thus behooves the beginner to practice listening to as many foreign stations as possible.

When getting ready to work a DX station, be sure to find out what is going on! It's infuriating to hear the questions on frequency from those who just won't take the time to carefully listen and put together what is happening. Is the DX station operating transceive? Is he operating split? Is he calling by districts? Is he identifying himself and giving out QSL information? The way to answer these questions is to *listen*.

One other thing about listening is that all frequencies should be scanned. It is too often assumed that DX *always* hangs out at the low end. This is not true. One friend of mine who is a particularly good tuner always turns up goodies in the most unlikely spots. He does it over and over again, and I've learned to appreciate the need for checking everywhere. Don't you forget it either.

Listening to CW is generally the same. Care and patience yield results. Spend enough time at each frequency stop to fully ascertain what is taking place there. Do not forget about CW. It is true that you will not be denied common countries by sticking only to SSB, because there is usually activity on both modes in most lands; however, it is not infrequent for CW buffs to put on rarer countries and operate only one mode, CW. K5VT has put on some real humdingers (9U and S9, for example), and he usually operates exclusively CW. Fairly rare Trindade (PY∅) was once activated by PY1MAG for over a month, and he was almost exclusively on CW. Thus a hard country was made easy by a long operation, but DXers had to get down into the CW bands to work it.

Calling the DX Station

Much of a beginner's initial work will be transceive with relatively easy stations, but DXpeditions and harder countries will be found and wanted. These stations frequently work split operation and use some system to break up the congestion. The most common method is going by districts in the U.S. and subdividing the rest of the world such as Europe, the Pacific, Japan by itself, Russians, etc. This type of operating procedure is usually obvious and is frequently expected by the masses. Newcomers should be aware of other systems for dividing up the callers, so that they can quickly adapt to the procedure that the DX station has selected. This will result in making fewer waves—and enemies—and increasing the chances of getting through.

When a pileup is very heavy, the DX station sometimes divides the U.S. districts even further. Usually this is done by state. This is particularly common on some nets that run lists. The Africana Net does this routinely and even specifies the number of stations from each state that will be taken. You might think someone can cheat and call when they're taking Texas when he's actually in New Mexico. There are several risks in doing this, however. Beginners frequently go back to these nets again and again because of the wealth of DX that can be worked this way. Sooner or later the net control will begin to realize that a particular call seems to be switching states. Also, even if one wasn't the loudest when his own state was called and thus he wasn't picked, his call may have still been *heard*. When it pops up shortly in the state next door, recognition is again possible. Though hams frequently get away from this form of cheating, one last revenge can be extracted. If suspicions are aroused, several DXers start checking the *Callbook* to learn the location of a given station. Or someone on frequency may know anyway. Right or wrong, the suspected culprit's location is publicly announced on the air, and I've heard QSOs scratched as a result.

Another way to subdivide the districts is to rotate prefixes. This system works by taking the 7s, for example, but then working each prefix separately. Usually the Ws go first, then the Ks, WBs and so forth. This system isn't very fair, because there may be more W7s than AB7s. Another disadvantage is the need to make sure that *all* the prefixes are called. On-the-air riots can start if one or more prefix groups are forgotten. One more disadvantage is that the method is inefficient. It takes an eternity to go through all of the prefixes and everyone's patience is taxed. Fortunately, this type of subdividing is being used less and less but one should be aware of it.

Another way to cut down the hoards is to take the first letter of the suffix. When I've seen this done, it's usually for the U.S. or even the whole world. This is often quite an experience. For example, if the DX station is on the letter B, this means K8B _____, JA1B _____, DL2B _____, WB5 _____, OE5B _____, and so on, may call. The pileup has a unique sound to it, because near and distant stations peak and fade when calling. It's very confusing and highly favorable to some stations while severely disfavoring others.

One other thing that might cause confusion to the newcomer is what I have dubbed the "minilist." In this interesting spectacle, the DX station has everybody call. He listens for several minutes while making a small list himself. When everybody finally quiets down, he begins to call some of the stations he was able to pick out. He repeats this cycle over and over again until he feels like shutting down. This technique produces some interesting negative effects. The pileup is huge, long, and obnoxious. Also, because one never knows if he's been heard and put on the list, each station has to scream without stopping until it's finished. This is really nice for those trying to operate nearby. While the DX station is working the list one by one, it's relatively quiet so several adjacent QSOs usually begin. Suddenly the list taking starts again, and this big ball of RF shatters the frequency and the new QSOs close at hand. It's quite cute. Though this technique is not common, it does not seem to be diminishing.

RULES OF BATTLE

> *Then shook the hills with thunder riven;*
> *Then rushed DXers, to battle driven;*
> *And louder than the bolts of heaven,*
> *"I QSL, you're five by seven . . ."*
>
> Cass

Sometimes there's a *lot* of thunder and clashing before that five and seven is reached. Though the linear, transmitter, and receiver may have no scars, the chief operator may appear to the uninformed as if he'd been out all night over New Year's Eve. Every serious DXer has been in pileups that have drawn blood—frequently his own. *If* the faces of DXers in the midst of active competition for a new one could be secretly photographed—heh, heh, heh—well, those would be *some* facial expressions.

If there is a great deal of roughness in a pileup, is there any sense in trying to organize "rules" for battling? Just listening to a mammoth pileup seems to confirm the opposite. There is no order, there is only disorder. There is no structure, there is only confusion. There are no rules of honor, there is only *might is right.* One thing is certain, experienced amateurs never show a potential ham a rough, aggressive pileup. It would certainly scare him away.

Yet, in this whirlwind of RF energy, there is order and there is technique. Just as a mating dance of an unusual species looks strange to us, the ceremonial activity is actually preset and perfect. A pileup is not much different. Even violators, who at first seem to upset what little order there is, are actually a ritualistic part of the scene. They are as much a component of the challenge as are the band conditions. The "policemen" have been with us (*against* us) for a long time.

There are four basic rules of battle that, if practiced, increase not only the efficiency of a pileup but also one's own chances of getting through. This latter aspect applies even if others are breaking the rules! Since getting through is ultimately what counts, these rules are not only benevolent, they are also wise.

Pausing

This is a true art. Though it is correct that for the DX station to hear a given call, a DXer has to transmit that call, it is equally true that one has to stop yelling sometime to see if he's been heard. Another fact is that while the DX station is transmitting, it is pointless for anyone to be calling; one simply cannot be heard at this time. This is so blatantly obvious and represents the very essence of common sense that it is surprising to hear the number of stations who do not follow this maxim. Yet there seems to be no end to those DXers who *never* develop the art of pausing. Frequent, carefully timed pauses are a must. Here's why.

Prolonged calling makes the DXer blind to what is happening on frequency. He misses the rhythm of the DX operator. He will lower his QSO success. If it's a transceive operation, he will interfere with others. He might antagonize the DX station who will not count him. Finally, he might antagonize others who will QRM him in return. There is simply *no* benefit to prolonged calling, because it actually reduces your chance in getting through. I have even heard—not infrequently—people still calling while the DX station is answering them. These operators actually miss their own QSOs, winding up not in the log. This is truly amazing.

Frequent duplicate calling is unavoidable. It happens to all of us, for while *we* are transmitting, we can't be certain that the DX station hasn't answered someone else. Outlandish, lengthy yelling is not the same and should be avoided. Here's how.

Develop the habit *early* in your DX career of letting up on the mic or key frequently. Concentrate on a style of transmitting that allows you to stop, listen quickly (with experience, you learn just how long), and resume transmitting in the same basic rhythm, only to pause again and again as necessary. As soon as you recognize that the DX station is talking, stop completely to learn what is happening. And if pause after pause results in never hearing the DX station, you must again stop altogether. It is possible that your transmit sequences are out of sync with the DX station's listening times. In other words, you're transmitting when the DX station is transmitting and listening when he's listening. A good pause will allow you to get back on track. Sometimes you find out that the DX station has signed; this is when you find out how silly it feels to be verbose in calling.

Remember that it pays to pause. Use the pause technique even if others do not. It does not matter if they don't get through because of poor operating habits. And remember that pausing is not a natural, simple act. It must be timed and planned carefully so as to allow maximum, rhythmic transmitting in addition to effective monitoring during the pause itself. Simply doing the act, however, will soon result in the skill of performing it well.

Combative Calling

Calling on top of one another is unavoidable in pileups, whether transceive or split. In fact that is what a pileup is—raucous, simultaneous, rib-kicking, competitive transmitting. It is not dishonorable to try to overcome your DX colleagues. In fact, if you get through you have performed an honorable act. However, there is a pileup phenomenon in which trying to out-do your competition can prolong the pileup length, ruining it for others *and*—like failing to pause—reducing *your* QSO rate. Here is what happens.

If a DXer is operating transceive or is occasionally monitoring his transmit frequency in split operation and he hears several loud competitors, a sort of leapfrogging calling duel may develop. Each one begins transmitting just as each thinks the other is abut to quit. Depending on how long the competitors push this, a pileup is lengthened unnecessarily. (Exception: in the minilist discussed previously, the DX station forces the unnecessary longer pileup.) In a regular pileup, however, certain DXers call as long as they hear anyone else. When the competitors are loud relative to each other they are likely to continue this mess even longer. Why is this done?

The primary reason for doing this is the belief that somehow it will enable the DX station to better hear one's call. The DXer thinks that this technique will either "fit" his call in between the others or outlast the others so that his is the final call heard. Does it work?

No. In the *majority* of instances, the DX station works one or more stations while the hard-heads duel on and on. This is true even in transceive, because there is always someone who can hear the DX fairly well and DX stations usually come back to someone who calls short, not long. One hears this over and over and over again. As the long winded battlers finally wither, the DX station is in the middle of his second or third QSO. It would be nice if DXers could learn this in the beginning as it would help make pileups more orderly. But since it's so common, let us spend a little time in trying to give as many hams as possible motivation for *not* doing this.

As stated, there are two reasons for employing this "technique": the belief that this will fit one's call in between others, in a sort of lull; and the belief that the very last call heard will be the one the DX station will pick up. In the first instance, this is the wrong way to play the "lull advantage." Though it will be fully discussed in later chapters, one finds the best way to sneak his call in during a brief quiet period in a pileup by *listening* not by excess transmitting. A carefully timed insert is always more effective. In the second instance, most DX stations do not pick up the last callers, particularly when three, four or ten or even more stations are carrying this out to ridiculous lengths. Almost always the DX station finds someone else and the big battlers get a contact much later, if at all. Even if the DX is inexperienced and taking the end callers, it is smarter to wait quietly and jump in at the right time when the others are getting tired and frustrated.

To sum up, it pays not to push the pileup out in time unnecessarily. Your QSO rate will only go down. This does not mean that I am advocating gentleness in a competitive pileup. *Au contraire!* I counsel aggressive calling. But be an assertive *thinker,* not an assertive *lummox.*

QSO Respect

Here we go again, folks—another obvious one. Once you recognize that another station has established a QSO with the DX station, stop calling! It is almost impossible to bluff a DX station into working you while he's working someone else. I have heard situations in which the DX had some call signs mixed up; here, the persistent caller may be able to get through. But in general, the DX station will never abandon a clear QSO for an intruder. In fact, more often than not persistent breakers only make him more determined to complete the QSO under way.

All of this is so obvious that stating it seems superfluous, however, I can't think of any other violation of ham radio common sense that exceeds the number of times that this concept is ignored. Genuine mistakes in timing your call can be made, of course, but experienced DXers know when someone is trying to deliberately get through on top of someone else. It seems that in the excitement some people cannot accept the fact that *somebody else* actually got through and, against all logic and courtesy, they believe that by continuing to yell, they will be welcomed by the DX station. Reality thwarts this notion time and again. Other than a genuinely confusing situation—when the DX station is uncertain of a call of which there may be several—there should be no continued calling. It doesn't pay off! In fact, it is quite common for the DX station to mark down a rude caller and not work him *or* work him but never send the QSL card.

So once again I propose motivation for decent operating based on the fact that not only does it help everyone else, but it helps you. Your QSO effectiveness will *improve* by not piggybacking. Overt rudeness will, in the long run, affect the QSO rate *negatively,* adding another minus to the always existing problems of band conditions, noise, and QRM.

Cop Out

Policemen—real ones—are fine. On the amateur bands, these vigilantes are a supreme nuisance. The comments I am about to make will hardly dent the thick skulls of those who deem themselves worthy of transmitting directions, instructions, criticisms, information, guidance, inspiration, curses, noises, bells, whistles, insults, and carriers, but I am going to put these people in their place.

For the very beginner, let us define the Amateur Radio **policeman**. *Ham cop: n. 1. Purveyor of advice and instruction on top of the transmitting DX station.*

It's easy to be a policeman. One needs only to have an Amateur Radio station and license. It's also easy to *want* to be a policeman. When one listens to a DX operation, the "numbscullery" that abounds stimulates us all with the desire to get on frequency and, with ultimate wisdom, utter a few words that will make everything right. As logic clearly tells us, this only adds to the confusion.

Who has the time to be a policeman? Those of us who need the new country are busy enough calling legitimately. There's no *time* to jump on the DX station and give orders. Thus, those voices that one hears playing cop are those who have *not gotten through.* Always remember this! The majority of policeman have made no contact with the desired station. It is at this point that we must make sure to include deliberate jammers (as opposed to those well meaning souls who *intend* to be righteous) in the same group. Therefore, in the final analysis there is no difference between policeman and jammers. The one thing that they have in common, and they hate to admit it, is that they *have not gotten through.* Thus, they have the time to sit on the DX frequency and make havoc.

There is nothing we can do about this sad group but relish in knowing that, for the most part, their frustration is real, their anger is pitiful, and their contact has not been made. There are a few—a very few—who have contacted the DX station and then hang around to play policemen. Usually, though, once a contact is made, the operator is so grateful to get through that any temptation to get on the DX frequency is lost. Thus, those interferers are almost exclusively hams without a DX contact. Take comfort in the fact that even if you didn't make it, neither did the cop.

To further motivate, promote, and even sanctify this concept, always remember that one cannot call the DX *and* simultaneously be a policeman or jammer. Therefore, no matter the frustration, indignation, horrendous anger and righteousness, *remain silent!* And call properly. For when it is all through, the *QSO rate*—that ultimately important factor—will be higher for those who have acted wisely rather than emotionally.

To the uninitiated, this advice may sound overemphasized. But it is not! The percentage of hams who, at one time or another, have played at least policeman (hopefully not deliberate jammers) is far too great. The problem is all too real and in such proportion so as to have prompted a very justified editorial comment in *QST.* And the newcomer, drawn as he might be to counter injustice by speaking out, should hold his tongue. The thrust of my comments is to show that despite the temptation to "do right," one will lose his objective: to *work* the DX.

GREAT CONCLUSION

Perhaps never have so simple a group of recommendations been put forth so forcefully. But, perhaps also, never has common sense been so flagrantly violated. The four basic concepts espoused here would do so much to clear up the confusion in DX operating that it is unbelievable. *Pausing, Clever Combative Calling, QSO Respect,* and *Copping Out* (not being a policeman or jammer) would make DXing almost easy. These four problem areas multiply the other detrimental factors such as poor band conditions, QRM and loud noise levels. But what's worse—what has been stated over and over again—is that the violation of these common sense concepts diminishes one's rate of contacts. If there is any message I could deliver to beginners, it would be to develop intelligent habits *early.*

3

Pileup Techniques

Of all the mysteries of DXing, the pileup is the most constant and yet different challenge of the competition. It is constant in that it is almost always a factor—in so far as hard to get ones are concerned—and it is different because those aspects which affect a pileup—DX location, band conditions, number of hams involved, style of the DX station—result in no two pileups being alike. In addition, depending on what part of the world one is in when hearing a pileup, its character changes considerably.

It would be interesting to visualize a pileup as an extraterrestrial observer would. Imagine floating in space, looking over the earth, and imagine further if the radio-frequency (RF) signals in a huge pileup could be illuminated like tracer bullets in the night. Just like long exposure photographs that show car lights streaking across the picture, one could see the glow of RF flashing across the earth and space. Signals ricocheting off the ionosphere, signals crisscrossing each other, signals struggling for dominance—all are part of this fascinating electronic scene.

And if one follows the flashes of energy, it soon becomes clear that there is a focus. While it is true that the RF radiates in several directions, it can quickly be observed that the concentration of streaking lights is zooming in one primary direction. It might be toward Kingman Reef or Heard Island or any other exotic DX locale, but our observer in space sees a blinding concentration of energy boring in on a single spot on the earth's surface.

While our visualization of RF energy as light is fantasy, in reality there is an invisible but ever real amount of tremendous radio electricity involved. This is all backed by—indeed it is all *caused* by—a tremendous amount of frantic human energy as desperate DXers almost burst at the seams with excitement. Altogether, in this amalgam of real and psychological energy, there is an enormous amount of unleashed power and force. And it is interesting that it goes unseen by most of the world.

From all this it is obvious that a heavy-duty pileup is a place of effort, emotion, exhaustion, *and* confusion. In fact, confusion is rampant. This is true even though it has been stated in previous pages that there is order in a pileup. Perhaps, it might be best to say that there is both confusion *and* order simultaneously. To further befuddle, it is also true that the degree of confusion is in proportion to one's expertise level. To rephrase, the "mix" of confusion and order is perceived differently by operators of differing skill. To the beginner, a simple pileup may represent the ultimate in chaos. This same situation would appear easy and orderly to an advanced DXer. Generally speaking, in a heavy pileup, there is much precision, skill and order while there is, undoubtedly, inherent confusion.

It is this—the struggle to overcome or minimize pileup confusion—that is the subject of this chapter. It is an awesome problem because there are thousands of signals intertwining with yours—all with the same objective. Also, many of the hams competing are not running small, humble rigs. They are using very high power. Thus, dancing successfully through this mess requires the grace of a runningback and the power of a defensive tackle.

PILEUP PERSPECTIVES

There are two basic aspects of a pileup. Both are interesting. One is from the DX station's point of view. The other is from the DXer's standpoint (your standpoint). Both should be understood by the DXer so he can appreciate what it's like at the "other end" and how to best cope with what's happening at "this end."

To the DX station, a major pileup is always awesome. Thousands of signals roar on a frequency or across several frequencies. As conditions shift, as people start and stop calling, waves of voices or code fade in and out. It is as if the ocean is constantly rushing after you, powerful and impersonal. And, as long as the demand is high, the clamor never ceases. It roars on and on, for even though the DX has satisfied many, there are always many, many more whose vital contact has not yet been made.

All this has an effect on the DX station operator (or operators). No matter what such a person may be like in normal life at this time of unusual concentrated mental activity there is a strain. It is simply impossible not be affected. Some people do have dispositions that enable them to adjust well. But many—and I mean *many*—undergo changes which directly and visibly affect their conduct on the air. These people may become, *in varying degrees,* snippy, catty, angry and tyrannical. When one is the DX one is in a high place and the urge to reign is supreme. Many of the worst offenders beg off by saying they are working under great stress and hazard, but this isn't true—they love it. And, yet, never forget that there is genuine fatigue on long operations with normal human lapses. The DX personality, however, aided by the strain, is frequently subject to whim and ego. All DXers know this.

It is important to understand that this is not an editorial to promote virtue in DX stations. People who live in remote countries or travel to obscure and hard-to-reach lands cannot be prompted to be anything but what they are. In fact, given the reality of being one of a few hams in a tough country or having traveled rough seas to a remote island, it's surprising at the number of real ladies and gentlemen we have. Iris (W6QL) and Lloyd (W6KG) Colvin, Father Moran (9N1MM), Tim Chen (BV2A/B) and Eric Sjolund (SMØAGD) come to mind quickly as exemplars of DX operating. In fact, Eric's ability to go to many rare DX locales, singlehandedly work long hours, and remain cool, QSO after QSO after QSO, is nothing short of remarkable. Of course, there are many others whom I have not mentioned here.

If these are "good guys," there are also those who, reacting to the pileup, lose a little, or a lot, of their control. Fortunately, really horrible examples are rare, but it is important to remember that the person at the other end is receiving the brunt of the RF, and it is difficult operating. Patience and cool will, not infrequently, be affected. The DXer should always understand and remember this and, then, develop a mentality that will handle it. The mentality is this: *calm. You remain calm.*

This sounds easy but it's not. When the DX is listening at 14,195 (kHz) and says he's listening up between 14,200 and 14,300—a spread that's so wide it makes a contact very difficult— the temptation to get angry at the DX is strong; however, resist letting it affect your operating. If you must, tell your buddies on VHF that the guy is foolish but then forget it. Get right back in action and pretend the DX operator is doing everything just fine. If the DX station skips your district, getting angry is easy. Do this later, however. In the meantime, remain calm, and wait for his next swing around. The important thing is to get the QSO! You can't ignore the mistakes or unfair things the DX does, but you must not let it affect you. Think for a moment about what it's like at the DX end. The pileup is like a cage and just outside are thousands of hungry wolves with sharp white teeth simply drooling to eat him up. Nothing you do can change his perceptions or his mind. So at least remain calm even if it's artificial. Just remember: get the job done first.

The pileup on the DXer's end is something else altogether. For one thing, the DXer only hears *part* of the pileup. Those stations that are close and with good skip dominate. Thus, the true depth and complete nature of the pileup cannot be fully appreciated. When the operation is split, the pileup doesn't affect the DXer much. Most of the listening is on the DX frequency. When one is listening on a group of split transmit frequencies, all one does is usually try to find a clear spot or use one or more of the jumping around techniques that will be described later in this chapter. When the operation is transceive, the DXer not only hears part of the pileup, but is affected by it. Obviously, the pileup interferes with receiving. There is nothing we can do about this, of course, but bear it. However, another thing happens that might affect the DXer's judgment. When one hears a loud competitor, it may be assumed that he's the station to beat. This may affect one's timing—wrongly. Remember, what the DX station hears is different: The "competitor" may not be competition at all. Hams close to you, however, will be in direct competition, but this is obvious. To be certain, *everyone* calling is competition. Just watch carefully what is going on and don't let a few loud stations trick you into abandoning a good rhythm of calling.

PILEUP RANGES

I suppose two hams are enough to technically make a pileup, but actually the range of small to large is more like ten, fifteen or so to thousands upon thousands. When first arriving at a pileup, it is important to assess the size but remember even only ten stations, particularly transceive, can sound like a lot. Don't let this trip you up and frighten you into starting up prematurely. Listen carefully for a moment and particularly between loud signals to see if there is depth to the pileup. If there is, your strategy will change of course. First make the diagnosis. Then begin the treatment.

Small pileups usually require only persistence, and—assuming you have a decent station and propagation isn't horribly against you—a contact will probably be made. Just relax and keep plugging away. Medium pileups are obviously going to be a little tougher and will require some strategy. Huge pileups are serious business, of course, and it is imperative to carefully figure out how the DX station is picking up people and to plan your actions accordingly.

Here's a little advanced tip regarding pileup size. If the station is rare and the pileup is small then, obviously, he's just gotten started. The operation will be briefly transceive, then frequently going split. At first the split will not be spread out much, so stay in a little close but check the transmit frequencies often. As the pileup grows, the DX station usually begins to spread out a little so move up in anticipation. On SSB, the DX operator will frequently announce his spread, but at first he will casually catch a few stations further up before he actually makes the announcement. On CW, the exact spread is usually *not* announced; instead the DX usually just states "up." Here the range of the pileup has to be closely judged. As it grows, spreads of 5 to 15 kHz are typical. The DX operator likes to encourage spreading out, hence much of his listening will be further away rather than just a few kHz up.

The range of pileups sizes is important. We'll rehash some of this again in the sections on strategy. For now just remember to assess what's going on in terms of the dimensions of the pileup.

EQUIPMENT

There is a full chapter, "The Station" (Chapter 9). In it, the design of a DXer's station and the technology of today's rigs are detailed. In a chapter on pileups, at least a few aspects of what kind of station is necessary in DX competition should be covered.

It's been said repeatedly in these pages that pileups are rough and unfriendly. Here the meek do not inherit the earth or anything—at least on a regular, consistent basis. We hear hams boasting all the time how they broke through a big pileup with 100 watts and a vertical antenna, but this just doesn't happen with consistency or ease. For casual DXing this is all right, of course. It doesn't matter too much if a Kingman or Mellish gets away. But for an avid DXer to miss one of these, it is disaster.

Hence, when thinking of DXing, when thinking of pileups and busting them, think in terms of having a serious station. DXers tinker and constantly improve their stations. It's time to develop this philosophy and start orienting yourself toward what's needed for *repeated* DX success.

Because of QRM and QRN (see Table 3-1), you should use the better receivers, transmitters or transceivers. A lot has happened. The state-of-the -art has changed drastically the last few years, and the latest generation of equipment provides many important advantages over older equipment. In fact, early solid-state gear is horribly deficient. Tube rigs, while all right, are no more than that. This is a good *time* to upgrade. The developments that have been made represent, for the most part, state-of-the-art to possibly the year 2000, according to at least one engineer (Ulrich Rohde, DJ2LR).

QRP (a maximum of 10 watts input or 5 watts output) DXing may be fun to some, but it's also masochistic. Serious DXers don't often care about QRPers and will rarely stand by to let one through at the cost of one of their own contacts. It's constantly bruising for QRPers. In addition, some DXers consider QRP to be anything *below a thousand watts*. The fact is that a linear is essential to DXing despite those who claim to be successful "barefoot" (without a linear).

A DXer's antenna system need not be a monster, but it should be as effective as possible. This is true especially for *receiving*, even though many people think in terms of the big signal on transmit. For 40 and 80 meters, do not use anything too fancy. Regular DXers who are on these bands are only trying to work Five-Band DXCC. After achieving this, they don't usually specialize on these frequencies any more. Higher frequencies require rotatable, directional beam antennas.

In essence, a DXer's station should involve, at every level, an attempt to have superior equipment. It doesn't have to be the best and certainly does not have to represent the most expensive of the choices available, but it should be better than average. After all, DXers are above average.

TECHNIQUE-WHAT ABOUT IT?

The techniques for handling pileups require knowledge and skills. All of us can turn the VFO knob and move around in split-frequency operation or transmit from time to time in transceive operations, but this is not technique. Technique is specific talent which, unlike sports in which everyone cannot be excellent, can be mastered by anyone who tries.

Table 3-1

Q Signals

Given below are a number of Q signals whose meanings most often need to be expressed with brevity and clearness in amateur work. (Q abbreviations take the form of questions only when each is sent followed by a question mark.)

QRG   Will you tell me my exact frequency (or that of . . .)? Your exact frequency (or that of . . .) is . . . kHz.

QRH   Does my frequency vary? Your frequency varies.

QRI   How is the tone of my transmission? The tone of your transmission is . . . (1. Good; 2. Variable; 3. Bad).

QRK   What is the intelligibility of my signals (or those of . . .)? The intelligibility of your signals (or those of . . .) is . . . (1. Bad; 2. Poor; 3. Fair; 4. Good; 5. Excellent).

QRL   Are you busy? I am busy (or I am busy with . . .). Please do not interfere.

QRM   Is my transmission being interfered with? Your transmission is being interfered with . . . (1. Nil; 2. Slightly; 3. Moderately; 4. Severely; 5. Extremely.)

QRN   Are you troubled by static? I am troubled by static . . . (1-5 as under QRM).

QRO   Shall I increase power? Increase power.

QRP   Shall I decrease power? Decrease power.

QRQ   Shall I send faster? Send faster (. . . wpm).

QRS   Shall I send more slowly? Send more slowly (. . . wpm).

QRT   Shall I stop sending? Stop sending.

QRU   Have you anything for me? I have nothing for you.

QRV   Are you ready? I am ready.

QRW   Shall I inform . . . that you are calling him on . . . kHz? Please inform . . . that I am calling . . . kHz.

QRX   When will you call me again? I will call you again at . . . hours (on . . . kHz).

QRY   What is my turn? Your turn is numbered . . .

QRZ   Who is calling me? You are being called by . . . (on . . . kHz).

QSA   What is the strength of my signals (or those of . . .)? The strength of your signals (or those of . . .) is . . . (1. Scarcely perceptible; 2. Weak; 3. Fairly good; 4. Good; 5. Very good).

QSB   Are my signals fading? Your signals are fading.

QSD   Are my signals mutilated? Your signals are mutilated.

QSG   Shall I send . . . messages at a time.

QSK   Can you hear me between your signals and if so can I break in on your transmission? I can hear you between my signals; break in on my transmission.

QSL   Can you acknowledge receipt? I am acknowledging receipt.

QSM   Shall I repeat the last message which I sent you, or some previous message? Repeat the last message which you sent me [or message(s) number(s) . . .]

QSN   Did you hear me (or . . .) on . . . kHz? I did hear you (or . . .) on . . . kHz.

QSO	Can you communicate with . . . direct or by relay? I can communicate with . . . direct (or by relay through . . .).
QSP	Will you relay to . . .? I will relay to . . .
QST	General call preceding a message addressed to all amateurs and ARRL members. This is in effect "CQ ARRL."
QSU	Shall I send or reply on this frequency (or on . . . kHz)? Send or reply on this frequency (or on . . . kHz).
QSV	Shall I send a series of Vs on this frequency (or . . . kHz)? Send a series of Vs on this frequency (or . . . kHz).
QSW	Will you send on this frequency (or on . . . kHz)? I am going to send on this frequency (or on . . . kHz).
QSX	Will you listen to . . . on . . . kHz? I am listening to . . . on . . . kHz.
QSY	Shall I change to transmission on another frequency? (Change to transmission on another frequency (or on . . . kHz).
QSZ	Shall I send each word or group more than once? Send each word or group twice (or . . . times).
QTA	Shall I cancel message number . . .? Cancel message number . . .
QTB	Do you agree with my counting of words? I do not agree with your counting of words. I will repeat the first letter or digit of each word of group.
QTC	How many messages have you to send? I have . . . messages for you (or for . . .).
QTH	What is your location? My location is . . .
QTR	What is the correct time? The time is . . .

This latter point is important to understand. DXing skills can be developed by almost anyone. I stress this. Of course, one has to be intelligent, sharp, good looking, confident, assertive and rich (how else will you afford all that equipment I just talked about?). All that aside, I wish to reemphasize that DXing skills can be learned. But it is just as important to recognize that technique *is* involved. For the most part ham radio is relaxing and casual—just flip on a few knobs and yak away. DXing, however, requires much more and the sum of it is technique—smart technique. Beginners, don't let this frighten you. Whatever else you are—banker, nurse, lawyer, shoemaker, metallurgist or Serbo-Croation translator—you have learned many skills. DXing is just one more. But think about technique as you read this book. Think of DXing as a specific skill and begin now to *intend* to develop it.

MENTAL ATTITUDE

The first order of technique is mental attitude. In sports, the need for a positive mental attitude is obvious. When this psychological lift is there, one can perform with coordination and talent above one's normal capability. It is less clear to me why this is important in DXing. It would seem—after one has installed a good station and knows the complete myriad of DXing tricks—that the rest is luck. DXing, in the final analysis, is an electrical hobby with a host of factors such as propagation and interference. After one's skills are fully developed, isn't it just chance that determines whether or not the DX station picks your signal out? After all, your mental attitude can't boost or forge your signal over others. Once your RF leaves the antenna and soars through space, it is beyond your control.

The above reasoning sounds logical; however, a proper mental attitude is essential. Just before the Heard Island DXpedition, the anticipation and talk were abundant on the high frequencies and 2-meter DX club repeaters. One day, I overheard two fellows in my DX club discussing Heard on 2 meters. One stated he needed it really bad (who didn't?) and he *hoped* he would get it. The other responded with to heck with *hope,* he *knew* he would get it. When Heard came on, guess which one got the contract first?

It is more than coincidence that some people get through to rare DX more often than others do. And while there are several reasons for success—big antennas and big linears, for example—mental attitude is also in there, right up front. In addition, the mental attitude required is different for beginners, intermediates, and experienced DXers.

Attitude at Different Levels of Experience

Probably the biggest stumbling block to getting beginners started in serious DXing is *fear.* Newcomers don't sit in front of the rig and bite their nails and wince; no, instead they don't even try when they hear a big pileup. Part of the reason may be genuine dislike for DXing, but pure hatred of DXing is rare. Most hams are intrigued by it, teased by it, and are titillatingly burned by it. This is evident by the number of casual DXers that can be found everywhere.

The newcomer arrives on the ham radio scene and views DXing at a variety of levels. Easy DX contacts are made, but those exciting pileups appear from time to time. The big ones sound so serious and complex that our initiate avoids them.

Those of you in this category need to readjust your mental attitude. In this case, it is true that there is nothing to fear but fear itself, because beginners, even with their flubs, are always welcome in the DX fold. Of course, if in the process of getting DX experience you mistakenly interfere with a DXer, he'll never forget it. Now, now, that's only a joke. On the more serious side, it can't be overly stressed that fear—especially secret, unrecognized fear—is no reason for not enjoying the complete thrill of DXing.

And what a thrill it is! When you get the commitment, a "new one" becomes vitally important. And while some ham critics say this is excessive, just remember that it is the same old *thrill of victory* that propels all competitive activities, professional or amateur. I want to use this motivation as one way to break down any degree of fear that might be preventing newcomers from getting an early start in DXing. Get in there! Start counting them! And suddenly you will start *caring*. Once this happens there is an automatic propulsion forward. Successes build and excitement takes hold. The need to get more countries rules. Then suddenly you're in it, competing devastatingly with a friend one moment, sympathetic with him the next if he didn't get through, congratulating each other if you both got it, and finally settling down later satisfied and jubilant with a sense of great accomplishment. But you have to get involved to savor it. Do not let pileups turn you off.

Intermediate-level hams have a more complex job in developing the proper mental attitude. These hams avoid serious DXing while secretly wanting to be a DXer. I am not talking about avid intermediate DXers, and for that matter advanced ones also, who have a separate mental attitude process that will be discussed next. Nevertheless, those people who have been around a while, are attracted to DXing, and yet still avoid it have a different and difficult problem. These hams want to be DXers but they aren't: they have the fear that beginners have, but they are very, very reluctant to *admit* it. This causes difficulties.

A large number of general-interest hams who have been around a while want to be DXers but are not. They do one of two things: they either state that they hate DXing (which they do not) or claim that they *are* DXers (but their country totals never seem to change much). Both of these groups are intimidated by the fierce competitive nature of serious DXing but they have a difficult time admitting it, even to themselves. It is not easy to reach this group for, unlike beginners, their minds are closed.

For those who *really* don't like DX, this doesn't apply. Yet, there are so many around who profess that they are in all those pileups when they aren't, exude that they can't wait for the next Abu Ail expedition but don't really try for it, or profess disdain for DXing that the number of would-be or closet DXers is obviously high. How do we reach them?

The only thing I can think of here is to get involved with some DXers, let this book motivate you, and force yourself to experiment on the air. There is simply no excuse to settle down after 100, 150 or even 200 countries. The pileups do get more intense but just develop the proper mental attitude and you can work them all. You have to *think winning* before you can do it!

Intermediate and advanced DXers who are not faced with an intimidation problem still must develop a positive mental outlook. This is important because in long, drawn-out pileups, there is strong temptation to quit when you do not get through. Let us explore this in more depth.

Mental Attitude As a Force in DXing

It is too easy to skip over mental attitude as a positive force in DXing because many—in fact most—pileups are not that demanding. With just a little effort, the simpler ones can be mastered. Many of the tougher ones are easy, too, because of luck or some skill or a combination of both. When a country comes on that is in high demand and that will probably not be activated again for some time, however, there will be high attendance. Almost every true-blue DXer will be there. Many casual DXers who hear about it will make an appearance. And many who don't know the first thing about serious DXing will bump into it and also attend. In short, the pileups will be simply enormous.

It is in these competitive situations that you will be extremely busy with pileup maneuvering and, if you do not get through, it is tempting at some point to lessen the effort or quit altogether. This may not be the first, second or even third day, but it doesn't matter when you quit if you have not gotten through. It's all over. The need to be persistent, to hang in there, and to keep performing at a high level of expertise, cannot be stressed enough. The following maxim is truth itself: *Almost all DXers with the right attitude get through to almost everything they try to get.* Some may want to eliminate the twice used word *almost*, making all DXers getting everything, but in reality I find that there is always something that someone missed, even if he gets it later.

The main point is, however, that with the right attitude the suc-

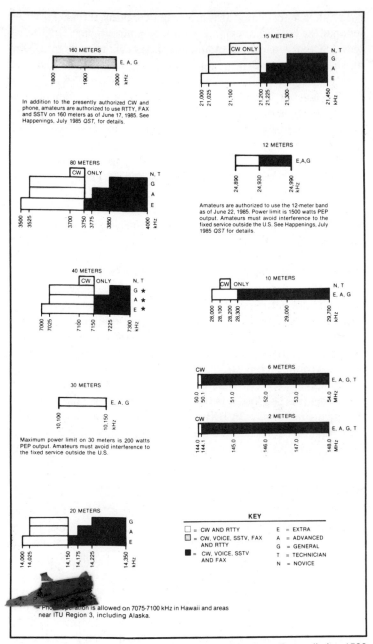

Fig. 3-1. U.S. amateur frequency and mode allocations. Power limits: 1500 watts PEP (200 watts PEP in the Novice segments and on 30 meters).

cess rate is astonishing. It is amazing how many serious DXers—without the aid of humongous stations—maintain high QSO rates with DX that has passed by in their time. Talk to any real DXer, and he will tell you that the right mental attitude is imperative. *Ya gotta believe!*

To develop this attitude, you must simply force yourself into a state of action, over and over again, despite mistakes and failures, until you become confident. At this point it becomes easier to get psyched up. Remember, though, that action or doing is the key idea. The knowledge of just *what* to do will also help tremendously. Never forget that even while learning you must still get in there and *do* it!

It is easy to say "do it," but it is not so easy to put this into practice. All of you, particularly those who have not done so before, should simply jump in the biggest pileup you can find and see what happens. Listen not only to the DX station but to those seeking to work him, particularly if it's a split operation. Feel the rush. Feel the excitement. Resolve mentally to become a part of it.

Don't forget the four common sense rules: pausing to listen carefully, not getting into a calling duel with someone, respecting a QSO that's underway and copping-out—i.e., don't be one of those airwave "policemen." And for goodness sake, relax. Don't take it too seriously. If you don't get through, there's no need to climb your tower and jump off.

Finally, however, if there is one serious aspect that you can take from this section, it is that like so many other things in life, attitude is important. In DXing, it may be surprising that it *is* so important, but at least now accept it. Then, in time, it will be easier first to develop and then to exploit your own positive mental attitude.

HITTING THE PILEUP

When a DXer runs across a good pileup, a few interesting things happen instantly. Though he may have been famished and eagerly awaiting dinner, the chief op is now no longer hungry. A favorite TV show may be coming on; this is suddenly of no importance. If the DXer is scheduled to leave the house, a fierce race is now under way to find out what the DX is and work it before he must go. As previously noted, other signs include an intense, anxious expression, profuse perspiration, and an increase in blood pressure. (No one ever said that DXing is healthy.) If one can't control his internal reactions on arriving at a pileup, at least he can control what should be done—or must be done—on the frequency for the sake of getting a contact.

Never forget that one of the most important aspects of pileup technique is the art of listening. So, when first arriving, listen! Find out who's there and what he is doing. You can't begin the treatment until you make the diagnosis. Experienced pileup watchers hear mistakes in this area over and over again. It's as if some people never learn. And, with one exception, one rarely increases his QSO success by yelling first and listening a distant second.

That exception is: If you land on a DX frequency and the DX station says he's on your call area or he's going to QRT after the next few contacts then—after you've at least determined if the DX is operating split or transceive—call immediately! This is the next thing to an actual emergency. Do take a second or two to think, first. That is, if the only country you have left to work is Burma and the signal you've come across is S-9 strength, then it's not very likely to be that. Don't call, therefore. If, on the other hand, you've got between 100 and 200 in the fist, then there is a reasonable chance that any given pileup may represent a new one for you. In this case, call quickly. It's just a case of using a little common sense before jumping in and transmitting.

Remember that this is the only exception when you should act first. In all other situations calling without establishing who the DX is and how he's handling the pileup will be counterproductive. Your chances for a QSO will actually decrease. Let us now examine precisely the different operating patterns of DX stations.

SSB

There are, of course, only two basic possibilities for all operations: transceive or split. However, within these two basic categories there is a great deal of complexity. As far as SSB is concerned, diagnosing a transceive operation is usually simple because the DX station and his callers are in one spot. Ascertaining that the operation is split is usually obvious also but not always so. For example, there are usually a number of people calling on the DX frequency by mistake. This automatically brings with it a certain number of policemen. Add to this the usual jammers, and a person arriving on the scene might think that the situation is transceive when it's actually split. The number of people who misdiagnose split operation is always amazing.

Another problem of diagnosis is finding the pileup but then being unable to locate the DX station. This is sort of like having the treatment but not having the patient. Many times the DX station is on the "standard frequencies" such as 21,295 kHz, listening up or 28,595, listening up. Occasionally, however, the splits are weird and it's disconcerting to hear this frenetic pileup between 21,300 and 21,330 and not be able to find the DX which is tucked away, almost in secret, at 21,245. Just remember that after finding a pileup, don't panic if you can't find the DX station. Scan further—and carefully—and the rewards shall be great. Note that for split-frequency operation, a remote VFO is essential.

Once it is clear that a given operation is split, the next task is to determine the frequencies being covered. This is usually but not always announced by the DX station. If not, listen to the spread and you can get a general idea of how wide the calling area is. One should also note that it is not unusual for the DX station to listen beyond the fringes of the frequencies he announces or that is apparent by listening to the pileup.

The next aspect that must be determined is the way the DX station is tuning around in the designated frequency zone. Put another way, we must find out how the DX is picking up the callers. Is he jumping frequency randomly? Is he moving slowly and sequentially up and down the range? Does he take tailenders? Is he concentrating on one or several frequencies in the zone even though he implies he's listening equally everywhere? Is his split unusual? For example, is he listening at intervals say of 5 kHz (21,305, 310, 315, etc.)? All of these questions are important, which gives the operator who takes the time to determine the DX station's pattern a substantial advantage.

I realize fully that when one comes across a pileup for a new one, the urge to get started is almost unbearable. But if the DX station is not on the verge of signing off, it is worth taking a little time to try to find out if there is a pattern to his listening. This is almost a necessity if the spread is wide (40 to 80 kHz, for example) as the likelihood of your transmitter matching up with his receiver becomes more remote.

It is difficult to say just how much time you should spend in finding out his pattern because it is also imperative to get in there and call. In addition, depending on one's experience and the degree of the mess, it may be difficult to determine any pattern whatsoever. At any rate, those who know how to diagnose the DX operator's habits and who spot the transmit frequencies are often there just ahead of the DX station. Such a DXer usually gets through on far fewer calls. If you find that simply calling isn't working, then I would counsel getting back to the pileup and making every effort to unravel its secrets.

CW

Many of the comments said about SSB are applicable to CW. There are, however, a few special things to consider. First, use a CW filter (500 Hz, for example). Even if you are employing very sharp filters, there always will be people directly on the transmit frequency who either fail to switch to separate VFOs or who are unaware that it's a split operation. Nothing can be done to prevent this, so it's up to you to listen carefully to determine if it's a transceive or split operation.

Another aspect that makes CW different from SSB is the announcement of where the DX station is listening if it is a split situation. The SSB DX station operator will usually say something like, "I'm listening between 28,610 and 630." On CW, however, the exact spread is usually *not* given. The vast majority of CW operators simply transmit the word *up* and nothing more. It is therefore up to the DXer to find the extent of the spread.

Some foreign operators do not announce even an "up" and careful tuning is required to determine what's happening. Other stations will even listen transceive occasionally and then start listening up from time to time. Such patterns are tricky and yet they must be discovered as one will simply not get through if one is calling in the wrong place.

Finally, though the majority of splits are up from the DX frequency, it is not rare for some DX station ops to listen *down*. This, too, is usually not announced and must be found by finding the people who are getting contacts. I have seen DX operators alternate, working one QSO up and then one down or a group of contacts up and a group down. Some of the BY operators listen up, down, and transceive.

The final word on all this is once again, *listen*. Before transmitting, spend some time finding out what's taking place. Specific steps for working the different types of transceive and split operations will be detailed in the coming sections but, right now, make a mental note to develop the habit of always checking the operation out first.

SINGLE SIDEBAND PILEUPS

Transceive. It would seem at first glance that transceive operation is simple. This is not to say that transceive pileups are easy but isn't it true that the technique is uncomplicated? Don't you just yell your call from time to time? The answer is a strong *no*. To the casual observer it may *appear* that no technique is involved and, in the chaos, "technique" may seem like a joke. But in reality there is plenty of finesse involved. In fact, heavy transceive pileups with the total mess on one frequency can be more demanding than even the wildest split operation.

It doesn't take long to get frustrated in transceive pileups, but step back in terms of slugging it out first versus thinking second. Nothing you can do will put order into the pileup or make anybody listen to what *you* think should be done. This means that the 5s who are calling when the DX is on the 4s will continue to do so. The desperate yellers who never stop to listen will call even longer and infinitely try your patience. And just when you think the DX has come back to you, that "Whiskey Eight" will smother the frequency with, "Did you come back to me?" As Rodney Dangerfield says, "No respect!"

It takes a cool head to ride through this turmoil but it is necessary. The first order of business is to get the contact and, as stressed before, let the losers lose their temper. The rattled and the jittery—no matter how righteous—finish, in fire and brimstone, *last*! Leave them alone in their misery. You sit there, tempted to preach but remaining silent, angry but in control, and you move steadfastly toward your goal.

To do this, you should first decipher what the rhythm of the DX station is. Timing is the ultimate key to transceive pileups, particularly when you are dealing with rare DX as opposed to everyday DX. How do you unlock this puzzle? I hate to say it, folks, but once again we must listen. Even if for a moment! It yields so much. I know I've said listen first to simply diagnose whether it's a transceive or split operation and what type of split is employed. Now, once again, after sorting out the primary question (where to transmit) comes further listening to determine *how* to transmit.

In transceive operation, where a semblance of organization and courtesy are routinely lacking, important factors are to be studied and exploited. The DX station usually has some type of personality that *makes* him handle a pileup a certain way. He may love the pileup and promote massive but friendly yelling *or* he may hate it and, in this unfortunate resentment, may cause friends to become enemies. Obviously, something between these two extremes is the norm. Whatever the thinking of the DX station turns out to be, however, it has better be sniffed out. For here lies the key to more efficient contacts after one has passed by the "easies."

Here are the basic response patterns of transceive DX stations. The rhythm of these patterns is the key factor.

- Listening, coming back quickly
- Listening, coming back slowly (call recognition important)
- Listening, making a "minilist"
- Listening, gets tired and "quits," only to come back in a certain time frame
- Listening, gets tired and *moves* to another frequency
- Really quits and signs off (point: don't believe it!)

Let us examine each in a little more depth. The DX station who comes back quickly is obviously picking out those hams who send their call quickly. It is very easy to get out of sync with such an operator and literally waste vocal and emotional energy in calling at the wrong time. It is important to say your call fast and, if you have a long call, perhaps only send part of it—such as the suffix. You should develop phonetics that come automatically to you for use in fast situations. If you find yourself out of rhythm, stop and recycle yourself.

If the DX is coming back slowly, remember that he may be having trouble recognizing call signs. Some DXers will use long cycles to send their call fast two or more times. However, remember that one carefully sent long call may help the DX station recognize all or part of your call sign. Another important thing to remember about slow responders is that the beginning sequence of a pileup is very intense. Because of this, some DX stations don't even listen initially, waiting for the energy to taper off. Then they pick out one clear call towards the end. If you find this happening, play your cards his way. Sit back quietly and patiently, let the vocalists get tired, and then pop your call in during the lull at the end. As with fast calling, you should develop phonetics that lend themselves to effective use in long situations. It may only be necessary to say your entire call more slowly, or the actual phonetics may change to increase length, such as with the letter S going from a short, staccato *Sugar* to a more leisurely, longer *Santiago*.

Those DX operators who take a minilist usually don't make it explicitly clear that they are doing so; however, it should quickly become apparent that this is happening as soon as he begins to work anywhere from five to 10 stations in sequence without taking any calls. The minilist was discussed in Chapter 2. Suffice it to say that though this is undesirable, the game must be played the DX station's way. To recap, the DX station listens to the roaring pileup and extracts a certain number of calls. He then begins to work those he's picked up list-style, one after another. Unfortunately, callers cannot be aware when or even if they have been picked up; thus, it's necessary to yell continuously until the DX says to stop. Ignore the inefficiency of this type of pileup and, if it's a new one, get in there and do it.

Another nuance is DX station fatigue or anger in which the operator says he's quitting only to turn back up again in some time frame, for example five to 10 minutes later. This is not the same as the legitimate QRT (station shutdown). This is an example of a temperamental DX station claiming he's tired, displeased with the pileup or—claiming nothing—suddenly ceasing operation. Equally suddenly he appears back on frequency as if he never left. Why this happens is the proper subject of a small treatise but there's no time for that here. It does occur. The point is that you should remain on frequency for a little while for the not infrequent return of the just disappeared DX.

Another maneuver made by some DX stations is to suddenly leave the pileup they've created and turn up on an adjacent frequency. This is *not* an unusual occurrence and should always be pursued if suddenly the original frequency becomes "thin." As soon as you suspect this might have happened, look around quickly. One is often rewarded with an easy contact. A final point: always recheck the original frequency in case the DX returns there.

Final trick: the guy who "really" signs off. There are a bunch of DX operators around who genuinely mean to QRT but for some reason never seem to be able to actually do it. To be sure, experienced DXpeditioners who are running a hot operation will QRT when they say so. But the large number of people who are casually QRV (ready to operate) in a somewhat rare country (embassy or government personnel, for example) are frequently not seasoned DX operators. As a result they often respond to the guy who begs, "Just one more QSO, please!" As abhorrent as this is, it works too often not to be mentioned.

There is one form of operation which is technically transceive but is, in spirit, the equivalent of split operation. This is the notorious *roulette* method in which everyone is calling in a spread of frequencies and the DX station chooses randomly, answering the lucky stations transceive. This type of operation violently disturbs other peaceful QSOs (as do split operations that are very wide) and deserves special comment. This will be described later.

Split Operation. On SSB, split operation has several general characteristics. Most split operations will listen up from the transmit frequency starting at 5 to 10 kHz higher. The average spread will usually be 10 to 30 kHz wide. The transmit frequencies are not infrequently on one or more of the three following standard spots: 14,195 kHz, 21,295 kHz and 28,595 kHz. Obviously, depending on QRM, the DX station might be just above or below these frequencies. The effect of the recently expanded 20-meter phone band on DXing should obviously be taken into account, too.

Of course, generalities are just that—the average—and DX operators love to be exceptions. Here are some situations which also are not uncommon. While the so-called standard frequencies are used often, *any* frequency may be chosen by a DX station. Never confine your 20-, 15-, and 10-meter search to the three frequencies mentioned. Always look elsewhere. In addition, while a general spread of 10 to 30 kHz is typical in a moderate pileup, larger spreads are quite possible. The majority of these larger spreads will cover about 50 kHz. Thus, someone transmitting at 14,195 kHz could be listening from 14,200 to 14,250. Common sense should say that this is already too much but, unfortunately, a minority like to go even further. Here we encounter the absolute wild spreads of 75 to even greater than 100 kHz thus knocking out almost the whole band. The DXer must be aware of such splits as usually on the higher end of the spread there is less crowding.

No matter what the spread actually is, the DX operator will pick up stations in that spread in a variety of ways. Some DX stations jump randomly around with no pattern or clues. Others will go sequentially up, down, or both, very slowly. Some will go up and down frequency but will do it in jumps. Still others look for clear spots and isolated signals. Some will sit on a spot and work tailenders. Finally, others will use a few odd combinations of the previous styles.

It should.be pointed out here that randomly jumping around is *not* the most common way a DX operator picks out calls. Usually there is some organized system or pattern in the way it's done. It is thus very important to learn these systems and exploit them. Many casual DXers ignore these advantages, and of course it's possible to get through without understanding what is going on.

However, smart tactics make the task easier and more efficient. And while a casual DXer can accept a missed contact, a serious DXer cannot. Understanding pileup strategy is one way DXers stay current with almost everything that's been on.

Random Listening. The first example of the way a DX station handles a split pileup is random listening. The name is self explanatory, and suffice it to say that the DX station bounces around on receive anywhere in his designated pileup spread. There is no order or logic about his tuning, and as such there is little one can do to anticipate where he'll be listening. There are a few options, however, which can help somewhat.

Let's say the spread is between 21,310 and 21,330 kHz. Obviously, the center frequency of 21,320 kHz is going to have to be passed by often. This is also obvious to everyone else, but a lot of people will call there. As a result, the concentration of transmitted signals will be high. However, you can follow the propagation of the DX as it goes up and down. When the signal is going through a peak, you can plant yourself on the middle frequency and have a better chance at getting through. Another thing that can be done is to choose a spot just up or down from the exact middle. Depending on the situation, this can lessen the amount of competitive RF.

The ends of the spread are another good option. *Many* DX chasers can't psychologically convince themselves that the DX operator really listens at the high and low edges of the spread. As a result, they are rooted with the masses on the center frequencies. Folks, the vast majority of DX operators *do* listen at the ends. Of course some don't, thus you should check the end areas and verify that the DX is picking up callers there. Once confirmed, use the ends, particularly if propagation is not favorable. If it seems that everyone is playing the end game, then slide toward the center slightly.

Another trick for random listening is to constantly check for the nearest thing to an empty or less-crowded frequency. These little "holes" move around as the pileup shifts to and fro. This happens because even though the DX station is listening randomly, good spotters still find the guy being worked. Everybody flocks to the frequencies around the QSO in progress, even though the DX operator will probably leave the area altogether as soon as this contract is over. All you have to do is find the quieter spots that result from this shifting, and call there. Remember that this is a very fluid phenomenon and requires constant switching to monitor transmit frequencies in the spread.

Finally, you can plant yourself on one spot and stick with it in the hopes that the DX will eventually find you. This isn't the most efficient method, but it's probably better than hopping around randomly like the DX station is doing. In the end, however, if *nothing* else has worked, you might try randomly switching transmit frequency. You may accidentally wind up in one of those less-crowded holes just as the DX station lands there. It's a longer shot, however, but when the DX operator is truly randomly switching, it's like doing a Rubik's cube blindfolded.

Sequential Changing. Probably the most common method that DX operators use is to slide up and down the frequency spread. This is commonly done by starting at one end, gradually moving to the other and then returning in the opposite direction. However, some will move in only one direction; i.e., they will go from one end to the other and simply jump back to the starting end of the spread.

The rate at which the DX operator moves is also very important. Some operators crawl across the spread very slowly and others move rapidly. There is usually no reason for the DX station's speed; it is simply his or her characteristic way of doing it. There is thus no way to guess how fast the DX station is moving. It simply has to be ferreted out. The only other thing that has to be understood is whether the DX is occasionally stopping and working tailenders one after another or if he's jumping over a few frequencies as he moves. These aspects are covered specifically in the next sections.

Of course, to unravel what the DX operator is doing, you have to take a moment to listen to the spread. Again, even though the urge to call is strong, it is smarter to figure out first the DX station's pattern. This is especially true in sequential frequency changing, because it is fairly easy to follow the listening trail. It's busy work, however, as you must constantly switch back and forth between the transmit frequencies and the DX station's frequency. It pays off though, more often than not.

To start out, you must abandon calling altogether and concentrate on the transmit frequencies. The first job is to find stations giving reports and see which way they are moving. Sounds easy, but it's not. It takes experience, particularly in large pileups. The thing to do is stick with it and determine the pattern. Pay particular attention to just how many kilohertz the DX station moves each time. This will determine his rate of change and allow you to more accurately plant yourself in the right spot on the frequency path he's covering. Try to avoid the temptation to jump on the same spot as one who's giving a report. While tailending this way does work, it's only effective if the DX is accepting tailenders. If the DX station is truly changing frequencies after each QSO, then calling on the QSO frequency will leave you one step behind all day long.

Remember, sequential up and down moving is probably the most common way DX operators listen in split operations. Once the pattern has been figured out, it is therefore relatively simple to plant yourself just ahead of the DX as he listens so that he lands on you on one of his moves up or down. It's tedious work to switch back and forth but it's worth it.

Leapfrogging. An interesting variation of sequential tuning might be called the hop, skip or jump. While it seems logical to *edge* up or down on frequency, there are DX stations who feel the urge to jump over the adjacent frequency next to where they just had a contact. This is a bit different from fast tuning. When a DX station tunes quickly, he still listens to relatively close frequency increments (say each 500 hertz on SSB) even if he moves one or two kilohertz before finding a clear signal. The jumper, on the other hand, literally skips over several callers to find what he thinks to be a clearer signal sooner. He's not altogether wrong since many callers have spotted the last QSO and jammed the adjacent frequencies.

The thing to do is monitor closely to see if the DX is shifting sequentially or in jumps. This takes more than a moment. It is thrilling to diagnose a simple upward or downward movement and anticipate where to call. However, if the DX operator is leapfrogging, and you anticipate the usual one or two kilohertz for advanced placement, you could once again be a continuous step behind. Take an extra amount of time to see if there is a consistent jump sequence to the DX station's movement. If it's always five or six or seven kilohertz, then draw the appropriate conclusions.

Never take the subtle difference between clear sequential and larger jump-like changes for granted. There is a logical tendency to conclude that a given operator who is listening in an up and /or down direction is covering all the frequencies. Some operators do. But many move in their *own pattern* with which *they* feel comfortable. If they're hopping, *you* hop. Otherwise, you'll be left behind. That's not a nice place to be in DXing.

Tailending. Nice name. It's difficult enough to find the end of many tails much less profit from them. Nevertheless, jumping on fast to a DXer's closing QSO is often very profitable. In DXing, tailending is common currency—something used every day. To those uncertain of this technique, suffice it to say you "follow" another's QSO—promptly.

Two very important things need to be understood about tailending: 1) If and how many times a DX operator picks up calls this way; 2) The need to let a QSO be completed. Since the latter is violated incessantly in tribal ritual, let us cover it first.

If the DX operator is taking tailenders, the overriding primordial instinct to call while a QSO is underway is very strong. Resist this pull! The DX station is, in the majority of instances, determined to complete a started QSO. Your call, at the wrong time, is more likely to aggravate the DX station, not endear him to your lovely call sign. DXers violate this principle over and over again, often when they've just arrived on frequency, and then find out shortly that a well-timed and polite call gets them the contact.

If you can keep this logical piece of advice in mind, then concentrate on finding out *if* the DX operator is actually catching tailenders. If he is, then jump on this situation at once, as it's likely to be fluid. There are a couple of things to coordinate. First, if the DX station is taking tailenders on a given frequency while *generally moving* up or down, this has to be noted! It doesn't do to tailend a dead frequency. Second, *if* the DX is staying put, find that out too. Sounds easy, but in the rush of events it's often not.

In the first instance, that of tailending when the DX station is also moving, it is imperative to judge how long he listens at one spot. Remember that this particular DX station is probably moving sequentially up or down as well. (You were told it's not easy!) So, if he is tailending *and* moving, you must find the average frequency shift and how long he stays on a frequency and accepts stations. Lots of wonderful switching back and forth is necessary but, strangely, this type of pattern is actually easy to follow because it's slow. There is the plus of small frequency changes combined with sitting on specific spots.

In the second instance, a DX operator doing "split" but taking tailending to extremes by sitting on one spot, there is an opportunity to be milked. Do it! Just remember an illusion is quickly created by the number of callers for whom a DX station is theoretically standing by. In other words, the number of callers increases quickly and this can create the impression that the DX station is still there when in fact he may have moved on. Therefore, call while this is going on but be suspicious and check carefully in the breaks. Be sure the station is really listening there. If so, be persistent and stick with this unusual opportunity. Persistent tailending, if accepted by the DX station, is a very effective way to get a QSO.

Tricks. Watch for little tricks that a DX station may pull. The DX operator may announce that he's listening between 28,605 and 625 kHz, but then pick out a few frequencies and switch between these. For example, he might rotate back and forth between 28,610 and 620 kHz. In the busy work of transmitting and monitoring the DX frequency, it's hard to catch this. Frequently, someone who already has a contact and is totally free to watch for the DX station's listening pattern will spot this and announce it on a DX repeater. Even if you're not in a DX club, there is usually a club's repeater in the vicinity. You should monitor this while trying to work a new one.

Another variation of this that I have witnessed was on one of the DXpeditions to Heard Island. The VK0 station was on 14,195 kHz and announced he was listening up at 14,215, plus or minus two. I didn't practice what I'm preaching here, and I called at 14,217 kHz on the premise it was likely to be less crowded. Had I listened to the transmit frequencies, I would have seen that even though he implied he was listening between 14,213 and 217 kHz, he was picking up almost everyone right on 215. Thus, my calling at 217 was wasted energy and I didn't get through. On another night this same operator announced a wider split but was favoring a few key frequencies over and over again. Once I spotted this, I made a contact quickly.

At first, this may be more than a little frustrating since what the DX announces and what he is doing are two different things. Let's explore this so as to better understand it and thus take advantage of such operation. When a DX operator is dealing with a huge pileup, he looks for the best way to get as many QSOs per minute as possible. By announcing a spread, he causes the pileup to distribute itself, thus reducing the concentration of RF and making individual signals easier to understand. If he announced he was favoring a few frequencies, everyone would crowd at these spots, thus eliminating the advantage of a spread out situation.

Is this fair? I can't really answer this since what a DX operator does is his own business, but you can benefit from this type of action. If you take the time and uncover that a DX station is doing selected listening, you have valuable knowledge. While many are yelling on the wrong frequencies, you know the right ones and this almost guarantees a contract. So, instead of railing against this in useless frustration, consider it to be part of the challenge. If the DX station is executing some version that you can unravel, you've got an easy one.

Unusual Splits. All DXers have, over the years, encountered a variety of ways that DX operators use to work a pileup. In split operations, particularly on SSB, instructions can be given by the DX station that result in different and unusual types of split control. These must be recognized and—oddly enough—must be believed by the DXer. The following example is representative of an unusual split situation and should be carefully noted.

On a DXpedition to Central Kiribati and American Phoenix, Erik, SM0 AGD, put this divided island on the air by switching boundaries every 24 hours. When I first heard what I'm about to describe, I wasn't so sure as, in the beginning, I thought he was transforming mole hills into mountains. Erik announced, clearly, that he was listening up in *5-kHz increments* from his transmit frequency. As I tuned across the spread, which was wide since he didn't specify an upper limit, I encountered an enormous pileup with peaks every 5 kHz. This seemed ridiculous at first. Not only was each peak pileup substantial, it was also difficult to know which spot he was on.

Fortunately, my frustration was quelled by a surge of common sense. All I had to do was stop calling and search the 5-kHz increments. Would you believe it? There was a pattern! What seemed infinitely complex was suddenly simple. Erik was moving first up then down in the 5-kHz jumps he promised and was working several stations on each landing. As soon as I found the spot where he was listening, all I had to do was to see if he was moving up or down in this particular sequence. Once the direction was determined, it was simple to place myself on the next step. I then worked him with two or three calls. The pileup was a major one. What Erik did was to make a hard situation easier for those who paid attention. His unusual but very specific split instructions enabled anyone who wasn't sleeping to take advantage of how he was handling his particular pileup.

This brings us to the next point on specific, unusual splits: be rational! Don't let variations of the norm rattle your concentration. Listen carefully to confirm the instructions and—excepting those who transmit obvious false information—accept them. Referring to my previous example, the information was clear. A DX station could, of course, give different instructions. The important thing is to not let a variance from the routine split cause confusion.

The next aspect is to have faith. Believe what the DXers says. Of course, the reader will recall that I pointed out that some DX operators will not do exactly as they say. I mentioned an example of a DX station favoring a few frequencies, even though he implied he was listening equally on the entire range of the split he announced. This is the exception. Usually, if a DX station says he is listening a certain way, he is. It's up to the DXer to verify this, of course.

If all this is correct, I ask why the following happened when Erik announced he was listening in 5-kHz increments? The largest pileups were every 5 kHz as expected, but there were also scores of stations calling in between. The number was significant, and there must have been great disappointments as, from what I could see, Erik pretty much stuck to the pattern as he had established it. Did these others simply not believe him? Did they misunderstand him? (He announced it this way: "I am listening 21,305, 10, 15, 20, and so on.") Whatever the reasons, a fairly rare pair of countries was missed by many who didn't pay close attention to what was going on. I think the conclusions are obvious. Heed them!

Roulette. In a technical sense, roulette DX operation is really transceive: Each contact is on a frequency with the DX station. However, because the QSOs occur on a spread of frequencies, it is very much like split operation and thus has been included here with other examples of split situations. For those unfamiliar with this type of DXing, here is a brief explanation. DXers wanting a contact with a given DX station spread out over a segment of the band and call. The DX station tunes around and answers various callers on their own frequencies. It is as unpredictable as it sounds. It is also an undesirable operating technique, because it causes havoc on a large portion of the band.

How does this happen? Since it's rather clumsy event, it would seem unlikely to even get started. I have heard of one DXpedition that operated for a time this way, but in general it derives from the frustration of an inexperienced operator in a rare country. Usually this station calls CQ and works a few people on a given frequency. Suddenly it's a big mess, and the DX moves from the pileup he created and calls CQ again. After a few clear contacts there is another pileup and another shift. DXers detect these movements, and a few people call on a clear spot anticipating where the DX might move. A clean, easy QSO is inviting to the DX station, and he comes back transceive to one of the callers. Things move fast from this point as others catch the clues. In short order, to the surprise of others in casual QSO, the band is churned up with a brigade of DXers strewn up and down frequency.

Quite correctly, the ARRL has listed this as an undesirable practice in its DX operating guidelines (reprinted later in this chapter). This is usually not a planned procedure and happens rather spontaneously. As a result it is something that most DXers face at least a few times in their climb up. It usually involves a difficult and needed country, so some find it difficult to walk away from it and not participate.

Once the spread has broadened, there is almost no pattern to the way the DX station selects his contacts. It should be remembered that this usually involves an inexperienced operator who is simply trying to have QSOs and avoid a pileup. The obvious thing to do is to call in the clearest spot you can find to facilitate having a good signal on the DX end. Of course, you can't be at all sure that there is not another QSO out of your skip range which is nevertheless coming into the DX station and interfering with your signal. Another problem is that if you are all alone on a clear spot, others will be quickly attracted and you will soon have competition. There is nothing that can be done about all of this of course, except persist and carefully select transmit spots.

Another thing to watch out for is a shot at tailending. Often the DX operator who is playing roulette will sit on a frequency and take a few tailenders, finally moving on when a semblance of a pileup appears. Because such a pileup usually occurs after as little as two or three contacts, one must spot and relocate fast to take advantage of tailending opportunities.

As with real roulette, a lot of luck is involved. In fact, I'd stick with Vegas odds versus the odds of getting through in DX roulette. The QSO rate is very low due to the inherent confusion, QRM and panic which ensues as desperate hunters seek very elusive game; however, if you combine skill with a rabbit's foot, it is surprising what can be accomplished. Many DXers who spot and tailend carefully and who persist instead of giving up do make it.

This wraps up SSB pileup strategy. Now to the low end of the band.

CW PILEUPS

Much of what has been said about sideband applies also to CW. There are some differences, however, and while running through the various situations, the concentration will be on how these differences affect the overall CW strategic approach.

Transceive. You will recall that with SSB, stress was placed on the importance of assessing the rhythm of the DX operator in picking out calls in a pileup (coming back quickly, coming back slowly because of the need for extra time to recognize calls, etc.) All of this is true in CW and need not be repeated here. The only thing that is lacking on CW as opposed to sideband is the minilist. Though I'm sure it can happen and probably has happened in the past, the minilist is, fortunately, extremely rare in CW operation. There are a few special things about CW, however, which can be looked for and which are helpful. The following is especially applicable in split operations but is covered here under transceive topics for simplicity since it does apply to both categories. A sideband DX operator frequently notifies the pileup that a QSO is over by saying, "QRZ." This is a little long in CW, and the CW DX operator instead usually signals the multitude that a QSO is over with a special signature. Probably the most common one is the short TU (for *thank you*) which is sent very rapidly in code. In particular, new DXers need to be aware that this is the "go flag" and that it's time start calling.

There are any number of little signals like this which a DX operator may chose or even make up. Another frequent one is a simple 73, sent once. Still another is a sequence of two dots, not too close as in the letter *i* but more like two *e*'s sent quickly, or the letter R (for "Roger"). To the newcomer, this may seem a bit mysterious but what else is new? Wait until you see what's ahead. At any rate, this "signature" comes up after many or most QSOs in some DX operations and should be easily recognized eventually.

One thing that's really important for the very beginner is to know that this signature is very often isolated or sent alone. In other words, while the DX may acknowledge the call sign of the last completed contact and follow it with the short code signal (e.g., *K5RSG TU*), there are at least as many and often more times when he does not. In these instances, there is only the solitary little clue transmitted by its little lonesome self. To the old timer this is as clear as blazing guns. The newcomer will need to concentrate some but this very quickly becomes second nature. Just be aware of it as you start out.

Another difference on CW is that your complete call usually should be sent. It will probably be remembered that on SSB there are instances when the speed of the operation sometimes requires that only a portion of one's call be transmitted (such as the suffix).

On sideband it's also not unusual for some operations (particularly lists) to ask for just one letter of your call, usually the last. This never happens on CW. Your complete call should be transmitted for clarity. Remember, alphabetical letters from multiple stations (with no phonetics and voice differences) easily get jumbled together in Morse code. A complete call may be easier for the DX to spot even if he misses or crosses up some letters. The DX op will more often get closer to your correct call if it's completely sent because of the logical sequence of prefix and suffix. Sending an isolated portion of a call can cause a mix-up with part of a prefix or suffix of others on the same frequency, in my opinion.

Two other aspects that are important about CW, particularly in transceive operation, is the code speed to use and the number of times to send your call. In general I would adopt the approximate speed of the DX station or go a little faster. A CW operator in a DX locale usually copies pretty well and his sending is often slower to assist less-experienced operators. If he sent as fast as he could copy there would be more repeats and confusion. It is thus safe for you to transmit a bit quicker unless you are dealing with an obvious novice. Though this is unusual in rare countries, it does happen. If someone is sending really slow, say between 10 to 15 words per minute, then by all means adjust your speed downward.

In regards to the number of times to send your call, remember the concept of getting into the rhythm of the DX station. If the pileup is not too large, this is particularly important as the DX operator is copying clear code—not voice with different and difficult accents, pronunciation and intonation—and thus his recognition time is usually faster. Accordingly, you should transmit a call twice or even only once in many situations. Naturally, there are other circumstances that require longer sequences of call sign transmission, such as when the pileup is large or the operator is inexperienced. With these exceptions, however, short sequences are standard.

Split Operation. CW split operation has some generalities that should be noted. Many operations, particularly DXpeditions, transmit on 025, such as 14,025, 21,025 and 28,025 kHz. Naturally, transmission is as near to these spots as QRM permits. In general, the majority of DX operators listen up with typical spreads of 5 to 15 kHz. As with the sideband "standard" frequencies and splits, the above information for CW covers a lot of DX situations and should be carefully noted by beginners. For split-frequencies of only a few kHz, your RIT (receiver incremental tuning) control comes into play on your transceiver. For wider splits, a remote VFO is essential. There are many exceptions, and you should never confine a search for a particular DX station to any limited group of frequencies. Also, all splits are not necessarily "standard," nor are they always up. I will now cover some interesting things that happen in CW operation.

As with sideband, there is almost always a pattern in which the DX station scans his spread (random listening, gradual up and down tuning, etc.) And also like sideband, the gradual sliding up and down the band is the most common method. Yet it is also true that certain stations favor tailending, a few key frequencies and the like. Different from sideband is the difficulty in spotting. One has to be pretty good with the code to keep one eye on the DX frequency and the other on the myriad of stations calling in the spread. It's very busy work and some discussion is necessary here.

If your CW skills are not sufficient to spot effectively, then you should abandon it. Spotting at all costs on sideband is effective because it's so often rewarding, but on CW—if your spotting is inefficient—you lose calling time. This becomes obvious quickly when you find yourself listening through more QSO cycles than you're calling. With careful spotting, some calling opportunities will be missed but don't get lulled into thinking that spotting is of paramount importance. If it's not producing results, discontinue it and concentrate on the DX signal.

You should also not abandon the pileup spread completely. It is a good idea to still check for clear spots in the spread from time to time. If there is a big concentration of signals on one or more frequencies, be sure to see if the DX station is taking tailenders or is favoring a few key frequencies.

Another thing to be done is to check your local DX repeater. In the really big DX situations, there is usually plenty happening on these repeaters. Quite often someone who has gotten through or doesn't need the country is concentrating solely on spotting. The information gleaned is usually announced on the repeater and you might as well benefit from it.

Another point to be stressed is the number of times to send your call. This was touched on in the discussion on transceive operation so a few comments will suffice here. In split operations, the DX op usually gets a rhythm going and, unlike transceive, he is able to stick to it. Transceive pileups frequently force a drawn out calling duel, but in split operations the signals are identified and answered quickly. It is imperative to recognize the length of the calling cycles and get in step.

One important point to remember is not to get into the habit of thinking the spread is only 10 kHz or so. This is particularly true because the exact frequencies of the spread may not be announced on CW like they are on sideband. Always check the spread for its limits. Never assume it is only a given amount wide.

Another problem more common on CW than on sideband is multiple DX operations with overlapping pileups. This was mentioned in Chapter 2, but it is worth another cautionary word as all DXers have witnessed the incredible mistakes—and missed QSOs—that occur in these circumstances.

Here's what happens. Usually two, but sometimes more, DX stations operate relatively close together. It is almost always chance and it is usually because of the tradition of getting close to 025, as mentioned. The stations usually indicate they are listening "up" and as a result the pileups intermix. The problem on the DX end is that some stations called are trying for the other DX station and therefore don't respond. *Your* problem is to properly identify the DX station *you* want. Remember that these stations often identify infrequently while concentrating on rapid exchange of call and report. Suppose you get the word that the VS5 you need is on the air right now. You turn on the rig and find someone who's giving out rapid reports, and you assume you're locked on. If, however, it's VK9 (Cocos Keeling) and you call—and make a contact—before the station has identified, you can come away empty.

Impossible you say! All you have to do is pay attention. That's right, just remember to do it! Dual or triple operations like this are not common, and there is tendency not to expect the problem. When it happens, though, it causes confusion and mistakes. The problem is that it's *not* extremely rare. It happens just often enough— and just infrequently enough—to trap hams every time. One example that comes to mind is when Heard Island was on. *Also* on nearby were China and South Shetland. China was easier to identify, but Heard and Shetland were both in the same general direction and operating typical DXpedition style: rapid QSOs, infrequent identification. The pileup spread was the same, so it was easy for both DX stations to pick up any given call. You can be sure that many people, not paying close attention, had a QSO with the wrong one.

The solution is simple and obvious. Spend the time to properly locate and correctly identify the station(s) you are trying to work. Never assume that the busy DX station passing out reports is necessarily the only rare one around or that it is the one you're seeking. A reasonably experienced CW operator never blows this, but many DXers operate predominantly sideband and come down on CW only for a new one. Don't you get stung!

Roulette on CW is less common and less offensive; in fact, true CW roulette is very, very rare. What does happen is a mild version in which DXers are aware that a particular country is known for coming up on a certain set of frequencies around a particular group of time options. Anticipating this, DXers call "in the blind" in the hopes that when the DX station turns on the rig he will hear said call. This works. I've already mentioned DX stations who are intimidated by large pileups and this enables such an operator to have a few clear and easy QSOs before the big rush. Usually, the DX station either stays put or moves around to just a few frequencies before settling down on one spot.

DXers should be aware of these operating habits, as anticipation and planning can result in an easy contact with a normally hard-to-get country. Personal on-the-air observation and scouting, DX bulletin information, and talks with other DXers provide ample data on which to build a strategic plan.

In summary, though certain aspects of competitive CW operation are like those of SSB, there are some elements which distinguish it. The experienced CW operator has no problem, of course, but most people are primarily sideband operators. New DXers in particular spend a very large percentage of their time on SSB soaking up new country after new country. Habits and techniques form quickly and infrequent excursions to the low end are sometimes handled in the same way as on sideband.

For the most part, a little thought and planning are all that's necessary for CW DXing. I'm far from an experienced CW man but I do all right because I've learned from the common sense directives I'm espousing here. I encourage all DXers to become involved in CW. It is very different from casual CW, of course, but there is one main reward: it is—compared to sideband—peaceful. It is much more gentlemanly, with a marked reduction in jammers, policemen and undesirables. When I first started DXing seriously, I preferred sideband. I now hail the operation that is all or mostly CW. New contacts are decidedly easier on this mode and —for you ulcer people—much easier on the nerves.

QRZ?

There is an interesting and frustrating thing that DX station operators occasionally do that should be covered. It involves both sideband and code operations. Here is what happens. After a QSO, the DX station, instead of signing or sending his indication for the pileup to go ahead, might do nothing. When the QSO is over, he simply tunes around for someone else. Because there is always someone calling in a large pileup, he finds a station. You suddenly hear him working someone else. There was no indication that the previous QSO ended. As a result, you didn't even call.

To make sure this is clear, let us briefly run through an example. Let's say you're on to a given rare one. DX says, "K9QQQ, you're 5-and-9." There is a pause while the K9 acknowledges and gives the DX his report. While waiting for the signal that the QSO is complete (a 73 or QRZ, for example) and calling should begin, you suddenly hear the DX op say, "WB8ZZZ, you're 5-and-9." This happens for a half dozen QSOs in a row. You haven't called. You haven't gotten through. You've just sat there. Suddenly a fuse blows—not in the rig but in your temper-tantrum circuit.

This failure to end a QSO can happen even with a good DX operator who is busy making QSOs and equally busy logging, but in such a case, it doesn't happen often; however, there are DX operators who make this their predominant style. Usually they do give a signal occasionally when they're starting to listen. When you've found one who omits this, however, what to do? You can't sit there all day and wait for the DX station to say QRZ every dozen QSOs. You can't chain and flog the DX operator like you want to. You *can* curse and pound your equipment, but this is often unrewarding. What do you do?

You have to get in there and call *but* it must be done right. Too often frustration sets in. The DXers call but that's it—all they do is simply *call,* without planning or timing. They say to themselves, "I'll show that _____, I'll call until he chokes on my RF." And call they do—over and over again, yelling into the mike or banging on the keyer, in anger and frustration. How do I know this happens? How do you think I know!

Folks, the thing to do is reset the calm button. Back off a moment and listen to the length of time a QSO cycle takes. Then you *estimate* when a QSO is over (perhaps even simulating the unheard transmission in your mind) and call at the approximately correct time. Don't over transmit here! Listen carefully for about how long it takes for the DX station to pick out a station and respond. If you're doing this right you should be hearing him come back fairly often. This is the way to do it. Refuse to allow yourself to fall into the easy trap of calling in frustration and therefore calling out of proper synchronization.

PILEUP WRAP-UP

Pileups are the meat and potatoes of DXing. Much time is spent with them or in them. Perhaps more than any other aspect of working DX, pileup management is the most crucial. It is the essence of competition. It is the area of least give-and-take in ham radio. And, if not tackled aggressively and wisely, the pileup will wash you aside like a riptide.

The usual counter to this is to think that an enormous linear and antenna will solve pileup problems. These do but only to a limited extent. If given a choice between having a great amplifier with a super antenna system, and having DX acumen (also with at least a table-top linear), I would choose the latter. The knowledge is worth at least 20 dB.

The other side of this coin, however, is that there are many DXers with smarts *and* huge amplifiers and antennas. Fortunately, the number with both are limited, and you can easily fit between the cannon fire. Ultimately, intelligent pileup handling is far more important than lavish stations. Few DXers have missed a country because they lacked very high power and large amounts of aluminum in the sky.

It should therefore be deduced that, given a minimally capable station, attitude and determination play a large role in DXing. This theme has been dealt with in some depth in this chapter and will appear again in upcoming chapters. *Belief* in the goal and the ability to attain it are essential. This belief does not come automatically, such as by simply *wanting* to be a DXer. It comes instead with a combination of planning, dedication, careful thought and, finally, equipment.

In the end, DXers—who deep down really hate pileups—are forced to live with them. *After* pileups are over, DXers might *claim* to love them (and to even have deep knowledge about mastering them) but in reality, in the moment of battle, when a contact has not yet been made, the pileup is a formidable enemy. All DXers have to contend with it. This brings us to the final point: The problem *is* the challenge. That's why we're doing it. Every challenge deserves a serious attempt at resolution. DXers know what their challenge is. Good DXers know how to meet that challenge.

DX OPERATING PROCEDURES AND GUIDELINES

By John Kanode, N4MM

(Originally published in September 1979 *QST*)

For many years DX operating procedures have consisted of split-frequency operation and transceive-type operation with good operator control. Recently, several other methods have emerged which, if not used properly, cause frustration, unhappiness and undue interference. These operations are known as lists, DX nets and roulette.

Described below are the different DX operating procedures and how and when to use them. Following these recommendations should increase everyone's pleasure in working DX, and in being DX operators, resulting in smooth, efficient operation which will allow as many amateurs as possible to make good DX contacts.

Split-Frequency Operation—Best Procedure of All

Split-frequency operation is, by far, the best method to use and results in less confusion, interference and frustration than other procedures. In split-frequency operation, the DX station transmits on one frequency and receives on another. This method gives the highest QSO rate, allowing more amateurs to contact the DX station. Anyone considering going on a DXpedition should plan to take proper equipment along so that split-frequency operation can be used. Any DX operator planning to purchase a transmitter/transceiver should purchase one that has split frequency capability or has the provisions for use of an external VFO.

It is recommended that a DX station operating phone transmit outside the U.S. phone subbands, when possible, and "listen up" (listen higher in frequency) for U.S. stations and "listen down" (listen lower in frequency) for other non-U.S. stations. Examples: *Phone*—DX station transmits on 14.190 MHz, listens for U.S. stations from 14.200 to 14.230 MHz and listens for others from 14.180 to 14.160 MHz. (Note: The latter represents pre-phone-band conventions, circa 1979.)

CW—DX station transmits on 14.020 MHz and listens up from 14.025 to 14.035 MHz.

No more bandwidth should be specified than the absolute minimum that permits rapid separation of the calling stations. Not everyone on the band is interested in calling the DX station, and the interests of others should be respected.

Transceive Operation—Requires Experience and Control

In this method of DX operation, the DX station transmits and receives on the same frequency. This takes less bandwidth than split-frequency operation, but more than operating split, its success depends on operator control and experience. There are three important points that lead to a successful transceive operation:

1) Experience and maintaining proper control;
2) A good to excellent signal;
3) Knowing when and how to divide a pileup.

Handling a pileup is an art and one must maintain firm control of the situation and know what to do to regain control should it be lost. There are some DX operators who can handle a pileup no matter how big it gets. But there are times when it is advantageous to divide the pileup in order to reduce it to manageable limits. There are many good methods now in use to divide a pileup. The most common method is by call districts.

Example: The DX station requests calls from the first U.S. call district. After working a specific number (usually three to 10), the DX station requests calls from the second U.S. call district, moving on to the other areas until he or she finishes with W0s. Then he requests non-U.S. stations, working a specific number of these.

Whatever sequence is used, it can be repeated as long as the DX station so desires. However, once a sequence is started, it should be continued until completed. This gives everyone a fair chance to have an opportunity to contact the DX station.

If the DX station's time is limited, the number of QSO's per call area can be reduced to, for example, three QSOs per district instead of 10. Whichever way the DX station decides to divide a pileup, he must be very firm and not QSO stations out of turn, even though some *will* call out of turn. To do so will encourage others to follow, leading to on-the-air bickering, and, in short order, chaos on the frequency. To prevent stations from calling out of turn, the DX station should announce often that he or she is dividing the pileup and the method being used to do so.

Another important factor in a successful transceive operation is that the DX station should have a good-to-excellent signal, so that calling stations will know who the DX station has acknowledged. Weak DX stations tend to be overcome by their own pileups and no one calling can be sure who has been acknowledged, if anyone. When this happens, QSO rates go down and confusion takes over. Then, the best approach for the DX station is silence until the pileup reduces itself to a low level or goes away. The DX operator regains control by announcing that he or she is going to divide the pileup and the method that he is going to use. Again, whichever method the DX operator uses, he must stick to it and give everyone a fair chance of obtaining a QSO with him.

DX Networks—More Orderly, But Slower

DX nets have been set up for many purposes. Some are for certain areas, club members, and so on. Some are for working DX only. They range from fully "open" nets for all to those controlled as to who can check in and when. For the most part, the DX operators who participate in these DX nets are permanent residents. They check in regularly and, with a reasonable effort, almost anyone can obtain a contact with them. If the DX station desires to participate in these DX nets, he or she has the right to do so.

Any net operated for the sole purpose of working DX should be operated in such a manner that all non-DX check-ins are given a fair chance to work the DX. Many of the guidelines used in transceive operation can be applied to DX net operations. Rare or semi-rare DX stations with limited time should refrain from using DX nets because of the resulting low QSO rate. DX nets are helpful to those DX operators who do not have the proper equipment to operate split frequency and/or the experience to operate transceive only.

Lists—Limited Use Only

List-style DX operations are, in some cases, the *only* way to work a DX station. However, it should be understood that lists should be used only when other DX operating procedures are not possible. Below are some guidelines that can make list operations run smoothly, be fair to almost everyone, and reduce interference to all.

Conditions under which lists should be used: (1) DX operator cannot speak English or understand it well or at all; (2) DX operator is *inexperienced*; (3) DX station has a very weak signal because of poor antenna or location, low power, or poor propagation.

List operation guidelines: Take a list for a DX station only if he asks or desires to do so. Do not pressure him. If it is determined that he is inexperienced in DX operating procedures, one could suggest a list operation, but leave the decision up to the DX station.

Take lists only for real-time operation. Old lists taken on a previous day or on different bands should not be used. Many on these lists do not show up, leading to frustration to those on frequency. It wastes time.

Prearranged lists should not be used. Lists that are not accessible to all who can hear the DX station should be avoided.

Only stations designated by the DX station to take lists should do so. When taking a list for a DX station, propagation differences should be taken into account. If necessary, the M.C. (master of ceremonies or list-taker) should direct one or two alternates in different areas to pick up the stations whom the M.C. cannot copy because of propagation. This gives all a fair chance to get on the list.

The M.C. should have good to excellent copy on the DX station, so that control can be maintained at all times. Lists should conform to the DX station's desires as to content, number per call area, length, and so on. It is much better to take a short list, about 20 stations or so, run them, then come back and pick up a second group of 20, repeating this procedure for as long as the DX station is willing to do so, than to take a list of 300 stations in one swoop.

Never relay signal reports and call sign corrections. If it is really a QSO, the DX station should be able to copy the call and report of the station he is working without the help of the M.C.

The M.C. should be clear and concise in giving instructions as to what he and/or the DX station desires. The instructions should be repeated often to reduce confusion and to inform those who have just arrived on frequency what is going on.

When working a DX station in a list operation, conform to the DX station's desires as to length of QSO. If the DX station is giving only signal reports, do likewise and do not give your name, QTH or a weather report. If the DX station gives reports and his name, be courteous and give him the report and your name; but in any case, conduct the QSO clearly and quickly, so that others can also have a chance to work him. Think of others.

Remember, list operations should be used only when required, not as a replacement for other unaided DX operating procedures.

Roulette—Should Not Be Used

At times, a DX station may appear on a predetermined frequency and announce that he will listen for callers within a certain band segment and will answer callers *on their own frequency*. This shotgun technique is referred to as roulette. This method is not at all recommended for DX operating, and should be highly discouraged. It causes confusion, encourages bootleggers, has a very poor QSO rate, results in questionable QSOs and produces a lot of unnecessary QRM.

The above suggestions are aimed at increasing your DX operating pleasure.

4

Band On the Run

It can be said that if pileups are our foes, then the bands—and knowledge of them—are our friends. Fortunately, there is a range of frequencies that offer different types of opportunities to work DX. And yet, these differences require some knowledge so we don't choose the wrong opportunity—when we have a choice. It will be pointed out that when BV2B is on 14,235 kHz, it doesn't matter what the hot band is at the moment. That's where he is and, regardless of conditions, this is where he has to be worked. True enough. But there are several hundred countries that we need that are not on one frequency and appear with varying degrees of regularity. The knowledgeable know *when* and *where* to look for them.

This is not to say that appearing on 14,221 at a prime time will guarantee a 4S7. But, if one knows that *long path* is open on 20 meters in the early morning hours and one gets up regularly to check it out, then one will find not only 4S7s but assorted VUs, A7s and an occasional A5, among many others. Turn your antenna north on 10 meters at the same time (with reasonable band conditions) and scores of rare Russians can be found—not in one morning, of course, but over a period of time.

Knowledge of what the bands have to offer is therefore valuable and worth studying. In addition, simple conclusions must be cast aside. Beginners realize, correctly, that 10-meter skip (during periods of good band conditions) is long and effective. What they must also realize is that it is very selective. Long skip does not at all mean guaranteed countries. Where the skip falls is just as important. It may be great that Australia is coming in five by nine. But that doesn't help much if you're looking for the 3C DXpedition and Africa is not coming in on that particular band. Further, as the sunspot cycle winds down, we shall find the sad deadening of some of our usually more favorable bands. "Skip" will be redefined within new realities.

In this sense, a book with a chapter on amateur bands has limits. Certainly, nothing about the conditions on the higher frequencies suggests band reliability—now or in the future. Even so, this author has traveled up and down two sunspot cycles (at eleven years each, no bargain), and my comments will reflect not only the look now but also the look for years to come.

These years, my friends, are bleak. Ten meters, a *great* band, has almost died. Fifteen is floundering. Twenty, as always, will survive, but barely. Forty shall thrive. Eighty and 160 shall also, particularly with its loyalists, and DXers shall overcome it all.

But! It's not all over yet. In fact, the combination of good band conditions far outweighs the sunspot doldrums and, for those who seek it, advantages are out there. Let's study them.

To do so, we shall look at each band as a separate entity which, as good hams know, each certainly is. They all have distinct personalities as it were, and unlike selecting a pretty girl or handsome boy, the bands, "ugly or good looking," are simply there to take or leave. Since DXers must take them, we might as well know something about them.

BASIC PROBLEMS

One thing that general-interest hams and DXers have in common is that both depend on the bands and band conditions together. However, the DXer is forced to utilize a particular band when a new one is on, regardless of good or bad conditions. Casual operators are more likely to QSY to a more active band when conditions are poor; however, there are more important reasons for knowing about band conditions.

For one thing, amateurs are active worldwide 24 hours a day. Knowledge of what area of the world comes in at what time and on what band enables the DXer to be at the right place at the right time. One can literally predict QSOs—even hard ones—with this vital knowledge. I can recall working A51PN, who showed up not infrequently in the early hours (U.S. time) on 20 meters. At this time, 20 meters is open long path; by pointing the antenna south (specific direction for a specific country of course) oodles of goodies from Asia and Middle East are available. In addition, since I live in the deep South, very often I had the jump on many distant points of the country because the ionospheric bounce into my area was stronger. (Others had propagation advantages to A5 on regular short path conditions at other times of course.) This knowledge—that the A5 frequently operated on this band at this time, that the long path was open and that my QTH was favorable—gave me a distinct edge in preparing for this QSO. And when the opportunity came, I was ready and waiting. Scores of DXers have stories like these. Much of the success revolves around wise use of the different frequency allocations.

Another time when knowledge about these things pays off is on DXpeditions. You'll read about multi-band operations all the time in DX bulletins, and usually a group is on more than one band simultaneously. It's possible, however, that you will only be able to copy them on one band at a given time. Therefore, the DX stations themselves pick a frequency or frequencies that are likely to bring them a lot of QSOs. They *know* about band openings, and the frequencies they choose reflect this knowledge. If you know too, then this makes the hunt for them easier.

Is this important? You bet. For one thing, start up times for a DXpedition are rarely punctual. There are almost always delays (days and weeks), and hours of searching can be saved by intelligent scanning based on knowledge of the bands. In addition, when a DXpedition does get going, those who are there first get easy contacts. All DXers can recount tales of very tough operations that were a cinch to work when they first came on.

The things to consider may sound a bit complex to the uninitiated. For the majority of the ham community, however, much of what will be discussed is just an extension of existing knowledge. In addition, there's not all that much to actually "know" or hold in the memory bank. A good deal of the work involved is getting on the air and, utilizing basic information, determine for yourself what happens and when on the different frequencies.

It's important to do it yourself because the bands change and there are no rules that hold. That opening on 10 meters that brought you that easy Moldavian contact in midwinter won't be the same next summer or even the next day. A solar flare can wreck K index and solar flux values, sending a band into an instant tailspin.

The knowledge that specific information about these events is broadcast on WWV at scheduled times can help you plan your strategy.

The DXer needs to be aware of several fundamentals. A working knowledge of propagation is essential. Whole books have been written about this subject. Here, only the basics of the basic are included. There are hams for whom this is a hobby in itself. DXers, however, don't play with propagation as an end unto itself but as means to another end: to get a new one.

It's important to know about band openings and closings. These differ on each band and, as mentioned previously, vary with the time of year and may vary daily. It is interesting that active hams develop a feel for these changes, and in a sense there is a comfortable comradeship with a group of frequencies. It's like knowing your own back yard.

DXers also need to be aware that propagation shifts differ on each band. There is a specific time when, for example, Africa booms in on 20 meters.

It follows that the peaks and valleys of band conditions have different time spans on the different bands. "Windows" on 10 meters are different than those on 20 and partly depend on what type of opening is taking place. In addition, some bands are more "brittle" than others, exhibiting rapid changes that are a problem for the DXer. All of us can remember a QSO dropping out suddenly. It's always a bit frustrating. But when you've tied up hours to get on a list and wait your turn, only to have this happen before your contact with FB8WG, then it's more than frustrating. It's catastrophic, perhaps exceeded only by something drastic such as the fall of the United States.

Another very important aspect that needs to be understood is whether the signal is coming through on short or long path. This may seem to be a simple matter to resolve, but the dominance of short path signals lulls even good operators into not checking for long path possibilities. Also, when long path is open, signals are often copyable on short path (though very much weaker), thus stacking the deck against even checking out long path. This is not an unusual occurrence. The tendency to point the antenna in the "logical" direction may eliminate the chance to find the rarer but significant long path opening.

The next consideration is the sunspot cycle. It marches along at its own pace and affects all of the considerations just mentioned. Thus, there is a great interplay of factors which affect what we are able to accomplish on the various frequencies. In a sense it is very much like sailboat racing with changing winds and variable seas which taunt and challenge racing sailors. It would be like playing chess with a shifting chessboard and moving pieces.

All of this might sound overly dramatic to the casual operator. To the DXer it is not. The competition aspect changes what would in ordinary circumstances be rather unimportant to something that is vital. Going back to sailing will help us draw a parallel. On a casual sail, if the wind shifts, one simply changes tack. But in a race, if there is a wind shift and one or more boats by luck happen to be in a more favorable position relative to the racing course mark, then those affected are ecstatic. Conversely, those not in a favorable position begin to use intense language common to sailors for centuries. A minor event—a change in wind direction—has had a major impact on a competitive situation.

In ham radio, it might go like this. Conditions into the Pacific could be favoring the west coast. An east coast ham, looking for a relaxed DX ragchew, might forego trying to beat the 6s and 7s for a VK contact and turn his antenna east for a chat with a G station instead. But, if an east coast DXer is trying to get T2 for a new one and the west coast boys have a propagation edge, it's a different story. After getting stomped for about half an hour or so, real frustration and anger set in. Where before was a calm, relaxed man or woman is now a smitten, resentful person capable of anything. It is stimuli such as this that result in us reading in the newspaper: *Enraged man dumps two thousand pounds of garbage on neighbor's porch.* There might be a picture of a perplexed woman quoted as saying, "I don't know what got into him. He just kept coming over with all this garbage saying something about those damn west coast kilowatts. Very strange."

Average readers of such stories pass these people off as crazy, but always there's a clue that tips us off. In the article, the police are quoted as saying, "This very angry man seems to have a communication problem." *We* know what that means. He couldn't communicate with T2. Explanation understood.

Don't let this happen to you. Beware of frustration and subsequent weird actions. Do your DXing with smarts. When you get angry, you can do something terribly intelligent like selling your rig and buying a set of golf clubs, or worse, full season jogging equipment. (About a week of this, however, will send you quickly back to the safety of your cozy shack.)

Well, we need all the help we can get. What our frustrated eastern DXer might have done to help him compete against the westerners was to turn his antenna *east* and see if there was a long-path opening to Tuvalu. Admittedly rare, they *do* occur and simply being astute enough to check for the possibility is all that's needed. Here are some specifics that should be considered from band to band.

FACTORS TO BE CONSIDERED

Opening and closing times for the different bands have already been mentioned. Suffice it to say here that these need to be understood in order to utilize a frequency. On a band like 80 meters, for example, specific openings to far away areas can be very brief—15 minutes is not unusual. Knowing this allows good DXers to pop in at these selected times and grab some excellent contacts.

Besides the general opening and closing times and selected brief openings (windows) of a band, there is also a broad peak time in which a given band is at its hottest. Being aware of these time spans can help you plan strategy. It is related not only to the particular frequency, but also to your QTH and the DX QTH. The reliability of these peaks (and accompanying valleys) is remarkable. For example, in my location, when 15 and 20 meters peak (and this is for hours at a time in the right months), much of Africa simply booms in at a certain time. If a new one is on from this area, the struggle to get through is frustrating; however, I know if the DX stays on long enough, as we emerge into our peak, my time will come. It has happened often, and simply knowing what the band does at my QTH at different times helps me to maintain cool. In fact, while waiting when conditions are not good, you can spend a little extra time between calls to check how the DX is handling the pileup. This band peak applies to all locations, of course, and DXers know when they are and what areas of the world favor them.

Another thing to consider is which modes are popular on certain bands. On 10 meters, for example, there is more DX activity on sideband than on CW. On 40 the opposite tends to be true. If one were to ask which mode/bands are the most common for DXpeditions, the answer would be sideband and 20 and 15 meters. As we hit the real bottom of the sunspot cycle, 20 meters becomes the mainstay. In addition, as the cycle winds down, there are periods when 15 meters is strong and periods when it is weak. During these interim phases, which will change each year, DXpeditions will select frequencies for concentrated activity depending on which bands are best. This part of band watching should hardly pose a challenge (but it could affect antenna plans for the coming years).

Long and short paths were mentioned previously. They are important factors to be checked. In addition, a signal may not always peak in the exact direction it is supposed to (in either long or short path). Alternate directions should then be explored. This is sometimes dubbed *skew path*.

The final consideration is propagation in general. DXers don't need in depth knowledge of this subject but a general overview is helpful.

A PROPAGATION PRIMER

Propagation is the lifeblood of DXing and radio in general. It is a virtual scientific field in itself, with scores of texts and associated scientific study. In addition, it is closely monitored and reported, which makes following it without becoming deeply involved relatively easy for DXers. While many hams involve themselves seriously in propagation matters, most do not. DXers in particular are usually busy exploiting propagation, not analyzing it. Ac-

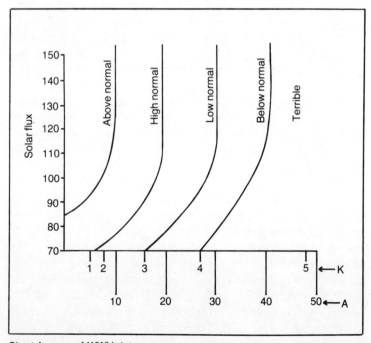

Chart for use of WWV data.

cordingly, the scope of the discussion of propagation will be limited to fundamentals.

There are three types of radio waves: ionospheric, tropospheric, and ground. DXing is accomplished with ionospheric radio waves, as a result of refraction (commonly called reflection) off the ionospheric layer. Actually, the ionosphere is composed of three layers—D, E and F (the latter of which is subdivided into two layers)—which range from 60 to 200 miles above Earth.

Ionospheric propagation depends on the thickness of these layers; the thickness is controlled by the degree of ionization. Greater ionization intensity is responsible for thicker layers and better bending or refraction of signals. In times of poor conditions, the layers are thinner and frequencies of shorter wavelengths (10 meters versus 40 meters, for example) pierce the less dense layer more easily. Consequently, these signals pass through the ionosphere without reflection and are received by hams in space. Longer wavelength signals such as 40-meter signals bounce back more easily off thin ionospheric layers; propagation remains acceptable. We thus see the extremes of a dead 10 meters and a live 80 meters with the spectrum of differences in between.

All hams have observed QSB (signal fading). This is a result of signals coming to the receive end on different paths. The "different" signals combine, forming a composite signal. This total or resulting signal has a strength which is affected positively or negatively by the phase relationships of the different components. Obviously then, variations of this interaction cause signals to move up and down on the S-meter.

Sunspot is no mystery word to most Amateur Radio operators. The sun is the immediate explainable source of radio propagation as a whole. Solar activity also has detrimental effects. Experienced hams know of marked ionospheric disturbance with sudden decreases in band conditions. A quick shutout of the upper frequencies is thus explainable when the sun "fails to cooperate."

Do we next take out our telescopes and observe solar flares? Hardly. Experts do this sort of thing for use and report it as well. With some basic knowledge, however, the data can be interpreted.

DX bulletins and the ham magazines are the main published sources. The ham magazines are limited by their advance publication setup to long-range forecasts that may not represent what is actually going on when the given month arrives. DX bulletins usually give two- or four-week predictions but add to this by effectively revising the information each week (that's how frequently they are published) and have the additional advantage of very late press time for input. Still, propagation is like the weather to a certain extent. Anyone who watches his local TV weatherman knows the limitations.

The phone-in opportunities seem to be more for propagation hobbyists than for DXers. First, you have to pay for the call. Then you have to deal with the data. This is fine, but for less effort and cost you can get adequate information from your receiver.

The ARRL sends propagation data with its W1AW bulletins. The WIAW schedule is published every other month in *QST*. Perhaps the easiest way to follow propagation is through WWV. At 18 minutes after each hour, the propagation bulletin is issued. It contains a review of immediate past conditions and predicted conditions. It also gives values for solar flux, A index, K index, and solar activity. DXers can therefore be rapidly informed about current band conditions. (For more information about WWV, see *The ARRL Handbook for the Radio Amateur*.)

You can do some interesting things if you get involved with propagation. Frequency and signal angle can be calculated, resulting in prediction of which frequencies are best for working a particular location. It is also possible to take into account propagation conditions and adjust antenna height for a particular distance. You need a readily adjustable tower, of course, and the mechanics are time-consuming. In both instances, the consequences could outweigh the benefits.

In the first case, calculating the maximum usable frequency (MUF), the answer may not equal the reality. For example, with a given set of conditions, the best spot to contact a TU might work out to be 17,400 kHz. But this is outside the amateur frequencies. He's more likely to be on 14,229 kHz, and you have to work him there. Similarly, by theoretically adjusting antenna height in certain calculated conditions might provide an edge; however, conditions might change, and concentrating on the antenna to this extent could cause you to miss something else important about the DX. Afficionados of DXing *and* propagation have their justified place in Amateur Radio. As a general rule, though, while DXing is subject to propagation, very little can be done about it.

Basic knowledge of propagation does allow strategic planning. If Glorioso is likely to be coming on this weekend, specific knowledge of how the bands work in that direction is useful. Why try 10 meters at the wrong time? Why overlook 40 meters because you don't like it when it might be the right one, particularly at this time in the sunspot cycle? Be wary of old habits that developed when the higher bands had different conditions.

160 METERS

Devotees of DXing on 160 meters, a.k.a. Top Band, tend to be viewed by others as gluttons for punishment. Why else would anyone stay up to all hours of the night (and early morning) just to get an earful of static? For those that see it that way, they should probably move back to the safer regions of 20 meters. But for those who really are up to the challenge, 160 meters is irresistible.

Top Band was at one time an exclusive club. It became known as the "gentleman's band" because some of the shenanigans often heard in 20-meter pileup just did not happen on 1.8 MHz. That "gentlemen's band" concept continues today, but the level of competition has increased manifold with the proliferation of transceivers with 160-meter capability. Participants are also discovering that acres of real estate aren't a necessity to compete. The two most common antennas found in use are a high inverted V, for those with towers of 80 feet or higher, and the inverted L, the favorite of the "tribander at 50-foot" crowd, who can locate the antenna next to the tower and extend the horizontal section to a support such as a tree.

DXing operating habits on 160 have also undergone some dramatic change with the burgeoning population and worldwide frequency allocations in a state of flux. The neophyte will soon discover, however, that North American stations do not initiate contact in the "DX window," 1825-1830 kHz. Calling CQ DX here, for example, is very much frowned upon. North American stations should, however, look for CW DX stations in "the window," where DX stations often operate split, announcing their listening frequency with "QSX _____." Transceive operation, both CW and phone, is commonplace between 1830 and 1850 kHz. The segment 1850 to 1855 kHz, is oftentimes used by Eastern European block countries who are restricted to that portion of the band. General phone operation is by convention; i.e, voluntary bandplan, recognized above 1855 kHz. The other band segment of significance is 1907.5 to 1912.5 kHz, the window for Japanese stations. JA stations will announce where they are listening, usually on the low end. VK and ZL stations will be found on CW at our sunrise, near the bottom of the band, just up from 1800 kHz, best during the northern hemisphere's winter season.

Top Band is strictly a nighttime DX band. Knowledge and use

of gray line sunset and sunrise information is probably of more use on this band than any other. The equinoxes can be especially productive. DXing on the Top Band offers a greater challenge than any other HF band. There is no better reward than to have another sleepless night produce a "new one." Where it once took years to accumulate 100 countries for Top Band DXCC, DXers are now doing it in a single operating season—meaning fall through spring.

80 METERS

Like the 160-meter band, 80 meters is, for some, an entity in itself. It's particularly important during the sunspot minimum. In addition, 80 meters almost never provides a DX opportunity that cannot be found on higher frequencies. Then what do DXers want from this band?

I've found that most DXers play around with 80 meters when they are chasing a specific award, particularly Five-Band DXCC. Once 5BDXCC is obtained, these hams tend to abandon the lower frequencies for the more traditional higher ones. There are others who—with justified reason and pride—concentrate on 80 for DXing. It's a challenging to work a lot of countries on 80.

From the standpoint of trying for Five-Band DXCC, you do not have to really kill yourself. Eighty meters is fairly reliable and its special times to work certain areas can be figured out and utilized. While super antennas for this band are difficult, this limitation makes the antenna problem easy and cheap: a simple dipole or inverted V can be very effective and competitive.

CW Operation

CW DXing on 80 meters is usually limited to the first 30 kHz of the CW band. Further, the rarer DX can really be found in the first 5 kHz. This is almost like a meeting place of friends on a park bench: reliable and constant.

This makes for easy tuning. You do not have to roam all over the place looking for the DX. This is perfect for the DXer who is only on 80 with the limited goal of Five-Band DXCC.

Most CW DX on 80 meters is transceive except for DXpeditions. In those cases, split operation may occur, but the separation is not wide. In addition, the DX usually hugs close to the low end.

If you are chasing the Five-Band award, concentrating on CW can be quite productive. Obviously, you are not going to go much over 100 countries, and for this limited objective, the CW yield is high. (For those concentrating on 80 as a main interest or serious subspecialty, tuning on sideband is necessary.) To emphasize the point of utilizing CW for 5BDXCC, I know one fellow who got his 100-plus-a-few in 34 days! He was using two antennas: an inverted-V and a ground-mounted vertical with no radials—hardly big stuff. This shows what can be done with the proper effort.

SSB Operation

Sideband DXing on 80 is a little different because different countries have different frequency allocations. Split operation in these cases is thus necessary, not optional. In general, sideband DXing is frequently concentrated between 3790 and 3800 kHz, the "DX Window." Note that owing to phone band expansion, amateur Extra class licensees may operate phone beginning at 3750 kHz, and Advanced class licensees may operate phone beginning at 3775 kHz.

Sideband operation on 80 meters is challenging. Though the signals might be weaker and the openings limited in duration, the substantial number of DXCC countries on 80 SSB make it worthwhile.

Noise

A constant and highly annoying problem on the low frequencies is noise. Listening to this noise is exasperating and exhausting. DX on 80 is often only marginal in terms of strength, and thus the noise is usually right in there with the signal—hence, the challenge. The special "sound" of 80 is one of the distinct aspects which marks the band.

HF DXers who concentrate on the higher frequencies often regard the sojourn to 80 as a prison sentence. One can hear the "moaning" on DX repeaters by DXers as they endure the 80 "punishment." Noise, noise, and more noise comes the complaint. It is almost an event when others suddenly realize that a particular DXer is no longer complaining about the 80-meter noise problem. All immediately realize that a safe margin over 100 has been worked. Thus, the consequential "noise" on the repeater ceases simultaneously.

Openings and Closings

DX work on 80 meters lives and dies by the specific openings and closings of the band. Whereas on 20 meters you can almost always find DX activity as long as the band is not totally dead,this is not true on 80. The latter can seem to be very lively. From the DX standpoint, it can sound like a cemetery. It is vital to understand this, as 80 meters cannot be adequately exploited without knowledge of its "personality."

First, it is essentially a night band. The DX opportunities come in usually in the late night to early morning hours. The type of opening will vary depending on your QTH. In the South, for example, there are evening openings to Central and South America. Occassional openings to Europe will appear. In the morning, about sunrise, 80 meters generally closes as a DX band.

Also, specific short windows crop up on 80. One example is the Pacific opening which frequently appears in the morning for a short period of time before the band closes. These particular windows can be as short as 15 minutes, and it is imperative to become familiar with their timing. Such windows provide different opportunities for the ham in Hawaii or the ham in Finland. It is up to the DXer to study the band relative to his area. For more information on 80-meter DXing, and DXing on other bands as well, see *The ARRL Operating Manual*.

Antennas

A detailed discussion of 80-meter antennas is beyond the scope of this book. For the first hundred countries, inverted Vs, slopers or verticals suffice admirably. For the serious 80-meter DXer, something else might be required. Those who have the commitment reap greater rewards as a gain antenna on this band is relatively rare. Those who take the trouble really stand out.

For the rest, it's more difficult. The antennas commonly employed provide no advantages as far as gain is concerned. Coming from the higher frequencies, where sophisticated antennas are the norm, an astute DXer may think he's in primitive back country. Used to begin a noble on 20, it's difficult to be a peasant on 80.

Still, these basic antennas are more than adequate. Remember, many hams leave this band after achieving a specific award goal. One thing that helps, though, is utilizing both a vertically polarized antenna and a horizontally polarized one. Our friend that worked his 100 countries in 34 days employed this method. With a good switching arrangement, you can quickly select which antenna is favorable at that time.

Courtesy

One of the big pluses for 80 meters is the friendly atmosphere and helpfulness almost absent on the higher frequencies. There is some pseudofriendliness that occurs in DX activity on 20 through 10, but on 80 meters it is genuine and is spearheaded by the regulars on this band.

This band is selective (and this applies even more to 160), which eliminates large crowds and the confusion that comes with them. This relative isolation allows for more gentlemanly conduct and a general clubhouse atmosphere. This more than helps to negate the noise problem. Eighty meters, with this atmosphere of courtesy, is usually considered a pleasure from the sportsmanship standpoint.

Schedules

The head bone is connected to the neck bone, and the 80 band is connected to the 20 band—for that matter, to the 15 band and the 10 band. Valuable sources for contacts on 80 meters are contacts with DX on the higher frequencies. When working DX on these upper frequencies, smart DXers inquire as to whether or not a station in a needed 80-meter country operates on that band. If possible, a schedule is frequently set up.

Far-fetched? Not at all. Remember, 80 is a special band. DX stations that might not want to make a sked for your best friend Mumphrey on 20 meters will still be delighted to meet *you* on 80. They appreciate the challenge of the lower frequencies, and if they operate down there they are usually willing to be most helpful. Making skeds is a most viable adjunct to 80-meter DX work.

40 METERS

The advantage of 40 meters is that the skip conditions are regular and reliable. DX opportunities might be greater than those on 80 meters. It's also a great starter band for beginners, because basic antenna requirements are simple. I can remember "living" on 40 for years. It's a great band, and for DXers, there are some great opportunities.

Basics

As with 80 meters, there is some simplicity for CW. For the most part, DX hangs around the low end: the first 30 kHz. CW operation is also like 80 with DX hanging out very low and separations that are not very wide. Sideband operation is virtually always split. DX stations transmit between 7040 and 7100 kHz, so check these frequencies closely. The DX listens 7150 and upward.

Openings and Closings

Forty-meter DX activity begins in the late afternoon or early evening and lasts until just after sunrise. There can be a lot of interesting windows. Those expecting an uninteresting band will be pleased to find long haul contacts with Australia, for example, to be commonplace at the appropriate opening. DX potential for rarer countries can thus be appreciated.

As with any band, you must use it to learn how it behaves in relation to your QTH. The problem for the casual user is to not let habits from the other bands dictate what you expect to find on 40 meters.

Antennas

Remember that beam antennas on 40 meters are *not* rare. True, they are not used as often as on the higher bands, but there are plenty of them. What's more, their owners are active. Those venturing to 40 meters for DX purposes will therefore find plenty of heavy competition.

The decision as to what type of antenna to use on 40 will be governed by your short-term and long-term plans for the band. If you are chasing a limited DX goal, I recommend a simple, inexpensive antenna. You may not get everything that comes on, but for 5BDXCC chasing, this is more than adequate.

There is one other point about the future, however, which modifies what was just said. As the sunspot cycle winds down, new DXers might want to consider heavy DXing on 40—not as a "40 lover" but as a serious DXer. Such a ham will not concentrate on 40 as a specialty but might want to have an edge for a new one when 10 and 15 meters are shot, and more DXpeditions turn up on the lower frequencies.

Miscellaneous Tidbits

The noise problem on 80 meters is less troublesome on 40. The summer noise can increase, but its constancy and the way it mixes with DX signals on 80 is different. Another hazard exists, though. Scores of shortwave broadcast stations operate on the 40-meter ham band. These stations occupy valuable frequency space. Shortwave stations are very powerful and simply must be accorded respect.

Forty meters is a night band for DX, but never forget distances of a thousand or so miles can be covered during the daytime. Thus, when first hunting for new countries, relatively nearby nations can obviously be worked with ease during the day. If you're in Hawaii, however, you might as well enjoy the beach.

The Future

For all years to come, the sunspot cycle will alter significantly what happens in Amateur Radio communication. The present sunspot minimum is no exception. Hams who have been through one know what will happen to a certain extent. However, many newer hams who are only used to good conditions will be surprised at the extent of the depth of the absolute minimum. Many DXers who have milked the peak conditions of the previous years will also be surprised; in fact, there *are* many hams who have gravitated to DXing because of the fantastic conditions providing a wide variety of DX opportunities in years past.

It is important to understand the value of 40 meters. Those who are locked into a 20-15-10-only mentality might miss a chance at some good DX. DX will continue to be the mainstay on the higher frequencies, but you should not totally forget about 40 meters. There will, however, be a difference of emphasis depending on one's DX level.

Remember that hams in many countries will be getting back on 40 now that conditions have simmered down. At certain times, a contact may be easier than on the higher bands. Finally, DXpeditions usually, though not always, field multiband operations. Eighty and 40 operation is becoming more common with the sunspot decline and will continue to be so. Of course, the 20-meter DXpedition station will still be on, but the other ops who would normally let loose on 15 and 10 (and who will so operate when these bands are open) will go to 40 and 80 quite often. To miss a new one because you're tuning 15 while the DXpedition is on 40 is sad indeed.

There are, of course, very different considerations for different level DXers. An old-timer with a 300-country total and a big antenna might do better to stick to 20 meters than go to 40 with a less competitive antenna. But the newcomer with 50 or 100 countries should regard 40 as an important band.

All of this is rather obvious at second glance, but I caution those used to those years of good conditions. Forty has been, for some DXers, unimportant. It will be easy to be lulled into overlooking it in the years ahead because of habit. If I had a lot of needed countries ahead of me, I would be planning a 40 strategy.

Keep in mind that a serious pileup on 40 is the same as a serious pileup on 15. The strategy for getting through is also the same. Forty meters in particular has begun to see larger, more difficult pileups. Because the windows are shorter, 40-meter regulars might do well to apply the techniques of competitive DXing. Conversely, those fresh from battle on the higher frequencies will have a "knowledge" edge on 40 that many low-key operators will not have.

Forty is a band of opportunity that should not be overlooked. I concede that the regulars on 40 meters will not joyously greet a large migration to "their" band. Though some hams will "come down," many will remain entrenched "up above." The smart DXer, however, is versatile and will definitely plan for 40.

20 METERS

Twenty meters is *the* all-around DX band. In good times it can be open for a full 24 hours with DX pouring in from various parts of the world. My log book during the last sunspot peak shows QSO after QSO—often uninterrupted—with DX contacts. One night the following three contacts were made sequentially on 20 meters: EP2TY (Iran), UA1PAL (Franz Josef Land) and ET3PG (Ethiopia). Two nights later: HZ1AB (Saudi Arabia), 3B8DA (Mauritius) and UJ8JCQ (Tadzhik). A few days later: 4S7DX (Sri Lanka), VK9XT (Christmas Island) and PP0 MAG (Trindade). On another night, in just 90 minutes, the following were logged: VK2AGT (Lord Howe Island), ZL2UW/C (Chatham Island), T3AT (Western Kiribati) and ZL2BCF/A (Campbell Island).

Some of these might not look all that juicy to old-timers, but remember these were logged one after the other in single, one-day (usually evening) operating spans. If I check out a few days in a row in the log here is another 20-meter yield: 9M6 (East Malyasia), A5 (Bhutan), CR9 (Macao), FB8X (Kerguelen), TN8 (Congo), and KC6 (Eastern Caroline Islands). These are pretty good catches on the way up, and they were logged in three consecutive days.

I don't mean to brag. This is just by way of example. This is—or was—20 meters. Many, many DXers have logs like this. Those were heady days. It was a thrill to see the awesome delivery capabilities of 20 and its remarkable reliability day after day. The world was at our doorstep, almost literally.

If 20 is *the* DX band, it stands to reason that it draws a big crowd. And its reliability produces another phenomenon more common to Wall street: investment. There is probably more money tied up in equipment and antennas for concentration on 20 meters than any other band. As a result there is plenty of RF running around here and pileups on 20 are legendary.

Openings and Closings

As previously stated, 20 meters can be a 24-hour band in good years. When we're near the bottom, it will be open from the morning to the evening hours and will often be dead at night. In the interim, there will be something in between these two extremes. This varies during the seasons with things being worse in the summer and better in the winter. Keep in mind that it's still possible to work 100 countries on 20 (and 15) and will always be so—even at the sunspot minimum.

What should not be forgotten is that there will be openings from time to time when 20 meters should really be asleep. These must not be forgotten. When 20 meters is open, very long-distance contacts will be possible, and considerable DX is workable at these unsuspected times. Finally, for the interim years, 20 meters will continue to be a very productive band.

Specific Windows

DXers used to selected, short openings to specific DX areas on higher bands will find this becoming more true on 20 as the sunspot cycle declines. When 20 is good, long openings to large areas will be available. But shorter windows must be gotten used to on the downswing, valley, and the beginning of the next uping. Of course, once we start going up again it's easier to readjust to longer and longer openings. and even though some have been through it before, the next peak—and peaks for years to come—will continue to surprise amateurs. And so will the valleys.

It is during the search for unusual windows that more attention to propagation must be paid. The windows of 80 and 40 are fairly predictable. The select openings of 20 will be less so during the decline. Checking with WWV, DX bulletins, and propagation sheets will be helpful. Remember that all is not the valley. Openings will be plentiful from time to time during the various years and need only be sought out. There must be an adjustment, however, to the fact the band simply will not be hot almost every night.

SSB and CW

In general, the low ends of both the sideband and CW portions of the band are where DX hangs out; however, the hangouts are not nearly so specific as on 80 and 40. Thus, it is not safe to confine searching for DX on the "usual" frequencies. DX can pop up anywhere.

Should one concentrate on sideband or CW on 20 meters? In general, both should be utilized. In the early stages of DX climbing, however, a greater proportion of basic countries (the first 100 to 200) is on SSB. Today, CW knowledge is limited in many countries. Thus, sideband is often the preferred mode. Ultimately, with searching, most of the basic 200 countries can be worked either on sideband or CW alone. A mix of the two, with emphasis on SSB in the early phases, is my recommendation.

Over the past years, however, the percentage of DXpeditions using CW has increased. This appears to be true in two aspects. First, the number of predominately sideband operations adding CW seems to be increasing. Second, pure CW DXpeditions also seem to be increasing. As I work my way higher, I find that I depend on CW more than in the past.

20's Peaks

One of the nice things about 20 is that while it is a "macho" band, this is primarily true in the evenings when *everybody* is home from work and on the air. The full 24-hour cycle of 20 is surprising for its delicacy and deliverance. Let us track an interesting and informative spree through a grand band. The following would not necessarily happen in one 24-hour period to one person but, if tried, it could.

Let's start in the early morning—say, about 1100 UTC. We get up. It's still dark but the smell of morning is present as we grab the paper for a brief look. With a cup of coffee in hand, we flick on the rig, watch its glow and listen to the sound of the band. It's long path time. Swing the antenna accordingly. Ahh, there's a VU, 5 and 8. Conditions are right. What's out there that we need? 4S7—Sri Lanka! Five and five. Do we need it? Check the worked sheet. Worked but not confirmed. Get with it! One call. Two calls. Three calls. "QSL the five and six. Name here is . . ."

Fine. Slip up frequency a little. A4. Nope—in the fist. Move on. What's this? A7? Quick check. Not worked. A new One! Linear switch! ?Where's the linear switch? Why isn't it turned on? Oh no! It's tuned up for 10 meters. Click, click, switch, switch. To heck with it, let's go. "A7XA, this is . . . !"

One minute, five minutes or possibly 30 minutes later after the QSO. "Yaaahooooo!" From a sleepy spouse in the distance: "Shut up! Sane people are trying to sleep!" Up the band a little more. No, back down. Something may have come up since. Tune, tune, tune. JY. Shucks, that's for beginners. Come on now, band: yield, yield! Here's a list. What's up? For what? For BV2B? BV2B!? Attack!

Later . . . whew! Two new ones. Well, it's good to be cool in a crisis. Almost time to eat. "Hey, honey, is breakfast ready?"

Now if one does not have to go to work, 20-meter operation continues further. Long path melts away. Other skips shift into place. It is QTH selective, of course, but relatively soon we hear Europe and its possibilities and the West with some goodies also. The usual southern skip brings in Central and South America. North brings mostly Canada and Alaska. The Russians are not usually available at this time. Asia and the Middle East are also not ready yet. But as the morning continues, some of Africa begins to open up a little.

One can free-lance on his own or work some of the list nets that are useful for beginning DXers. One might easily pick up a Midway or Johnston Island, for example. As the day wears on, the general north European area gets stronger. This could bring in a JX (Jan Mayen) or an OY (Faroe Islands). In between, chats with DXers from England, France, Germany, Italy and other European countries round out the time and maybe provide some DX info.

As the afternoon comes on strong, more shifts take place. Africa starts coming into its own. By midafternoon to late afternoon, the continent really booms in. One day a rare 5X (Uganda) came on. I was at home copying the 5X in the early afternoon before his peak. He was copyable but was rather marginal. The east coast boys had the edge. The South was generally getting clobbered in the pileup. The Midwest and West weren't even in the picture yet. A friend of mine who needed 5X was on the DX repeater traveling about town. He had about another hour to go before he would finish his work and get home. Of course, he was ready to burst at the seams. (I was cool—I already had 5X.)

We both knew what would happen. I reassured him that the 5X was still weak, that the "max" had not set in. I predicted that by the time he got home, the 5X would be in our backyard and he would have an easy contact. When he did get on the air, the 5X had come up three or four S units. He got an easy QSO. By now the 5s and southern 4s were milking the opportunity. That was 20 at it's best: delivering a rare one with accuracy and predictability.

Time moves on. As evening sets in, the whole world can present itself. Europe peaks even more and good skip to the south opens up. The Falklands might be found, and from time to time DXpeditions bring up some pretty good ones, such as South Sandwich and South Shetland. These openings last for hours and along with the Pacific, which also starts picking up, a veritable wealth of possibilities exist. The Middle East comes into its own and the rush for YI (Iraq), YK (Syria) and others is frequently on. The Russians are now coming in good and the pileups for the rarer ones are strong. In fact, everyone in the U.S. is home now and 20 meters is like the City of New York at night—fast and busy.

As the night continues so does the DX. It was in the early to late evening that Heard Island was prominent during the last DXpedition. Heard is a considerably distant location from the United States and I can remember the pre-expedition talk. "The windows will be short," many said. Comparisons of stations as near as possible to Heard seemed to verify this. Certain Antarctic stations had brief openings during the evening. An FB8 on Kerguelen very near Heard Island was on the air occasionally but wasn't on long enough at a given time to see if the opening was truly short or long. There was no definitive evidence of what was going to happen and there were almost as many predictions as DXers.

In the end, 20 delivered, even if not to everywhere equally. Heard was on for hours at a stretch over multiple evenings with a signal that was very readable. In addition, there was a late evening fadeout followed by, for those who stayed to observe, a comeback. In our area this comeback was almost guaranteed. Once it was discovered, it was only a matter of waiting while the Heard operator was gone or "in the mud" for him to rise up again. Those so waiting were often on the edge of the last skip with a propagation advantage. The contacts at this time were a snap.

As 20 goes past midnight and into the morning, the DX frequently continues. Pacific openings prosper and this is when A3s (Tonga), H4 (Solomon Island), ZK1 (Cook, North and South) and the like are nabbed. And so it goes. As we slip toward dawn, the long path begins to open. One last scan. Then its time to turn the rig off. Reluctantly: click. The LEDs and meter lights face. Now to crawl into bed. As your spouse grumbles at the disturbance, you exclaim with enthusiasm, "Twenty—wow! What a great band."

The reply is terse: "Shove it, lid." Ahh, well. I never said DXers are *perfect*.

Equipment

The concentration of power and antennas on 20 meters is remarkable. It is true that people who construct very large linears have bandswitch positions for other bands. But, as all DXers know, these liners are really for the 20-meter crush. It wouldn't be surprising to see *rust* on the 80- and 40-meter contacts of the bandswitch. The percentage of DXers that own the large commercial linears is probably higher than the percentage of hams in general. These linears are advertised as having the ability to put out cool kilowatts of RF and they do it. You don't think DXers would find these linears essential merely for 15- or 10-meter DXing do you? Of course not. They're purchased primarily for 20-meter use.

The same is true for antennas. Ten-meter and 15-meter arrays are much simpler to install and maintain. Considering the tower size, rotator size, overall installation, and maintenance required for large 20-meter antennas, one might expect the percentage of these arrays to decrease somewhat on this band. Not so for DXers! *This* is the no-compromise band. A friend of mine has a wide-spaced, five-element monobander for 20. Perched on top of this is a normal but comparatively tiny three-element duobander that he uses for 15 and 10. This should clarify for us where his priorities lie.

DXers, would-be-DXers, newcomers, and the potentially curious should understand that serious DXing requires a serious station—not the best, not the most expensive, but a *very* good one all the same. This is too often forgotten in the complexities of studying or considering DX on-the-air technique. Certainly there is no better place for such a reminder than in the section on 20 meters.

Long Path

A short word about long path. Earlier comments in this chapter raised the subject of long path. All amateurs are—or should be—familiar with long path propagation. The thing to do is not forget about it. There are times when it is, for all practical purposes, the only path. When the long opening is hot on 20 meters in the morning—say, to Asia—short path is practically nonexistent, at least in the deep South. Other similarities apply to other locations. In such circumstances, the poor ham who hears others calling and jumps into the fray with his antenna in the regular direction is lost. Ahh, but that's so obvious you say. True enough in certain situations when long path is exceptionally dominant. There are times, however, when this is not the case.

In these situations, signals may be heard on both paths. If the DXer is locked on to the thought that short path is right, he may miss a better opportunity. I remember trying to work 9M2FR on the PHO net. The list was taken and moving pretty fast. The northeast, midwest and west coast gang were getting their reports fine, but I was having a lot of trouble copying. From the various signal strengths of the DXers, it was obvious everyone was going short path. No one even mentioned long path.

I didn't know where the few other 5s on the list were, but as the fifth district was approached, I swung my beam quickly to long path. Out of the mist came the 9M2, three S units louder than on the short beat. Somewhere in the nation was a border where short path propagation diminished and long path took over. I was across it. I may have been the only one that night, assuming the other 5s on the list were west of me. And also assuming the 4s were not in the deep south. (Long path would have been superior for them, also.) At my QTH, this long path improvement is rare and, hearing the signal on short path, I almost tried it the wrong way. As conditions were, I would have missed the report and not had the contact.

The moral here is clear and obvious. Woe unto ye who forgeteth *pathus longus*.

15 METERS

If 20 is the band of muscle, 15 meters is the band of opportunity. It provides excellent selected openings in which signals from exotic DX locales can come in with substantial signal strengths. It has its times of reliable, repetitive skip openings as well. It is, however, more finicky, dropping out suddenly or peaking unexpectedly, or doing both. And when the sunspots put the *gri gri* on it, it's going to have long bleak moments of quiet.

All in all though, 15 meters is a band that offers good DX opportunities, is less crowded, requires a smaller antenna, and is generally pleasant to work. Several 15-meter nets run DX stations (discussed in more detail in the next chapter). Also, a lot of DX hangs around operating "freestyle" or free-lancing" (i.e., on their own without the aid of list-makers). Fifteen meters is just good enough to be an excellent DX band and just peculiar enough to keep the hordes off of it. Rare ones, however, will draw the usual big crowds. In the grand scheme of things, 15 was well planned.

Openings and Closings

During the good years of a sunspot cycle, 15 can be open almost 24 hours on some days. At the very least it opens early in the morning and closes late at night, so good DXing can be accomplished most of the day. As a given cycle winds down, the opening starts a little later and the closing moves much earlier. At the bottom, there are many "dead" days with only short periods of openings. During the off years, there will be intermittent specific openings from time to time.

Another unfortunate happening is the psychology of the down cycle affects the amount of use of the band. Hams assume that the frequencies are dead and don't get on. This is regrettable because there will be many times when the band is open and it simply won't be used because of the belief that it is dead. When a major contest comes on, this will be verified. Many will be surprised at the propagation, but after the contest most will retreat once again to the lower frequencies.

There is very little a DXer can do about all this except take some comfort that fair-weather DXers will, for the most part, drift away from the sport. This will contribute to lessening of pileup intensity, particularly on 15 (and 10 also) which dedicated DXers will continue to use. To utilize these bands wisely during the off years, it is a good idea to follow scheduled DXpeditions through the bulletins and from on-the-air information. In addition, WWV should be checked to see what conditions will be like. During certain openings there's a good chance for picking up some good ones with ease since pileups will be small.

Peak Times

During the good or average years on 15, the broad peaks deliver a remarkable amount of DX. It is different from 20 meters, however, and DXers should be aware of the characteristics of 15 meters. In general, the band is more selective than 20 when it is open. There are times on 20 when almost all parts of the world can be heard at once. On 15, the openings are usually to only certain areas. During an entire day, the shift rotates, bringing in new areas and losing others. During good sunspot years, these openings are predictable and fairly reliable but again not nearly as reliable as 20 meters. As the cycle winds down, it is this reliability that changes the most.

Only so much advice can be given about any of the bands and you will need to study the specific characteristics of what's open and to where at *your* QTH. Nevertheless, understanding each band's personality at the outset is helpful to this task.

General Comments

The CW and SSB comments discussed for 20 meters are, for all practical purposes, the same for 15 meters. I would add that sideband DXing is probably a little more prominent than CW DXing compared to 20, but again I stress the need for DXers to utilize both modes. Power and antenna requirements are less, but more often than not the push to provide the best on 20 results in having something to spare on 15. This simply provides more of an edge on this band. Once again, please, don't' forget to consider long path possibilities.

DX tends to be concentrated on the lower ends of both the CW and sideband portions of the band (like 20). The common SSB DXpedition frequency is 21,295 kHz. As always, nothing about DXing is fixed, so never assume this frequency is sacred. DX shows up when and where it wants to.

Fifteen is a band which has great potential and often delivers. Its strength is that even in the bottom of the cycle it will have openings. The challenge to DXers is to get on the band and to have some luck. The luck requires a run of three: first, the band is open; second, the DX is there; and third, the DXer is there too. This combination may seem rare—and indeed it will be at the very bottom—but for those astute enough to intelligently search, there will be DX so easy to work that stories of same will make 20-meter diehards go into spasms of agony!

10 METERS

Perhaps no other band has mood swings like 10 meters. When it's hot, baby, it's hot; when it's cold, it's like an arctic winter. Dramatic description? Yes, because that's the way it is. It is really almost depressing to go through the good years on 10 and then endure its dormancy. To tune kilohertz after kilohertz and hear absolutely no signals seems incredible. Those who have not experienced the hot years will simply think that this band has no value. Don't believe it!

When the 10-meter band peaks at the sunspot maximum, its long openings are almost as surprising as long closings of 20 at the bottom of the cycle. To hear 10 blossom is a pleasure that should not be denied the DXer. And to the DXer who tries, 10 meters shall return favors.

One of the special things about 10 is that here, finally, selective skip pays off. To a degree not found on the other bands, the DX usually skips in nicely (if it's the right time and place), and the competition is usually out of skip with you. Thus, the DX is generally stronger than the competition, and transceive operation, which is more common on 10, is easier. Even if stateside is loud on your end, rarely does it obliterate the DX like it does so often on 15 and 20 meters. This "private" connection with the DX is both reassuring and practical. Contacts are decidedly easier on 10 when it's open. If one could only compare the agonies of chasing and then working something on 20 with the relative ease of working the same country on 10, there would be no choice. It's like a comparative paradise!

Looking back, the Russian countries come quickly to mind. I nabbed many tough Russian-block countries with ragchewing ease on 10 meters in the morning hours. I can remember these same countries appearing on 20 in the evening with truly enormous pileups and low QSO rates. The irony was striking. If ever there was an editorial plea for even semi-smart DXing, this is it. How could so many people exert so much screaming energy for a tough 20-meter evening contact when all they had to do was wake up a little earlier in the morning? The 10-meter counterpart was ridiculous. Easy contacts, with a bit of ragchewing, were the *norm*.

For those who encounter 10 meters for the first time at the low point of the cycle, the potential for fantastic DX conditions must be accepted on faith. Those who have seen the good years and will experience the bad already have the faith. At any rate, 10 meters is a band that is ignored by much of the amateur population. Good DXers know its value. Newcomers should not forget about 10 meters even in its worst years.

Openings and Closings

Normally 10 meters opens in the morning and closes in the evening. During peak years it remains open into the dark hours for varying periods of times. Sometimes it will stay open very late. During the absolute minimum, the band will be dead for long distant DX work much of the time. There will, however, be openings from time to time.

During the last sunspot minimum, DXers in New Orleans had reliable contacts with Central America and the Caribbean. In DX contests, they we surprised to find that Europe was also coming in. That's a fairly long haul for the South. I was living in Germany during the last minimum. I watched it for over three years. Ten meters seemed better there, of course, because so many countries are compacted in one relatively small areas. France was just an hour's drive south from my QTH. Thus, it was possible to talk to many countries. The experience of 10 in those off years was different from that of stateside hams.

Another thing about 10 in *any* year is that the summer-winter differences are more dramatic. The band falls off a fair amount in the summer; when this occurs, the opening is later and the closing is much earlier. The reversal of this starts in the fall and goes through spring.

Peak Times and Windows

During the good sunspot years, broad peak times are available in the daylight and early evening hours. Remember that, like 15, skip is selective. This means that only some areas will come in at certain times. This selective aspect has one advantage in that the pileup is usually not that strong and you can hear the DX better. The counterpoint to this is that some areas may be heavily favored while others are disfavored or don't even hear the DX at all. This occurs on all DX bands, but can be quite marked on 10.

Another thing about 10 meters is that it is a more brittle band, particularly in the off years. An opening can open and close suddenly. This is particularly frustrating with list operations. You make the list and wait to work the DX, only to have conditions fall out from under you. Ten meters makes up for this with its selective openings.

Specific windows pop up on 10 at any time. In fact, during the off years, that's just about what propagation on 10 will be: a little series of intermittent windows that open and shut relatively quickly, as opposed to a dawn opening and dusk closing with plenty of good conditions in between. The DXer will have to familiarize himself with these specific windows and what opportunities they will offer.

General Comments

There is much more DX on sideband than on CW on the

10-meter band. By this I don't mean common foreign countries. If one tunes both modes one will hear "DX" coming in on both modes. In the search for new ones, however, there appears to be greater activity on sideband.

In the CW mode, DX tends to hang out at the lower end of the band (as on 20 and 15). Sideband, too, is concentrated on the lower rather than upper portions of the band; however, on 10 there is an important factor; the segment of frequencies is large. The phone portion starts at 28,300 kHz but rare DX is *common* to 28,650 (and some might argue further). The DK2OC DX Net, for example, operated on 28,750 kHz, and it was a check-in point for excellent DX. Though it takes time to tune this large frequency spread, never forget to do it. During the bottom of the sunspot cycle, there is probably more concentration on the lower ends to take advantage of the shorter openings and make finding one another easier. The "usual" DX frequency for DXpeditions is 28,595 kHz with a split at 28,600 or a little higher and as wide as the DX designates.

Compared to 20 meters, 10 is a much less crowded band. Adjacent frequency interference is less because of this. Almost everyone knows how frustrating it is to work weak DX on 20 with two guys running kilowatts just a few kilohertz away. Signals on 20 meters easily run 40 to 50 over S9. An s-2 DX signal suffers next to such high quantities of RF. The 10-meter band is much more pleasant in this regard.

A friend of mine is a real 10-meter buff. She has participated in just about all of the 10-meter activities and contests and chased most of her DX on 10. She has approximately 250 of her current 304 count from 10 meters. A friend of hers who also concentrates on 10 has even more countries. The point here is that 10 is a serious band with tremendous potential. I emphasize this again because its potential is ignored by many hams.

As we go through the current cycle and conditions deteriorate, many will forget about 10 meters; however, those who use it selectively on the down side and who utilize it more on its return up will catch some easy DX. The important thing is to stick with 10 to keep being familiar with it. When good conditions return, the DX will also return. Remember antennas and power requirements for 10 are less of a problem, and DX stations are attracted by this fact. Those familiar, ready and waiting will get significant, easy DX contacts which are being fiercely contested on the lower bands at the same time.

IN CLOSING

This discussion of the bands has brought forth some aspects of their individual traits. In some parts of this discussion, the information may have seemed obvious. Conversations with others tend to support the general idea that the bands have not been used widely as they could have or should have been used. Certainly my friend's 250-plus countries on 10 meters represents an example of what can be achieved outside the heavily populated frequencies. She reports more enjoyment of her contacts than I can remember with some of the same ones on 20 meters. This is ample testimony for breaking out of the mold. And remember that band/mode versatility is crucial in chasing Five-Band DXCC or Five-Band Worked All Zones (see Chapter 8 for a summary of these awards).

Another reason for emphasizing a multiple-band approach to DXing is where we are in time. We are in the midst of a sunspot minimum. This encourages some hams who like to work DX to concentrate on 20 meters. In this sense, this book is very appropriate now and, more importantly, for the sunspot cycles to come. The temptation to be lured away from productive bands—particularly on each side of the very bottom—is counter productive. For serious DXing, there is a commitment to search *all* the bands. It is—or should be—understood that this commitment requires an above-average effort. The pursuit of DX on all bands is part of that effort.

5

The Hunt

The hunt is the real essence of DXing. It is in this area that skills truly vary. And it is here where true DXers outdistance their casual counterparts. Many armaments for the hunt may be utilized by the DXer—some of them controversial—and the employment of these weapons gives serious DXers a substantial advantage over the rest. Those weapons do not include antennas or power or even pull, though pull does help a few. The fundamental element here is knowledge—deep knowledge of how DXing works, deep knowledge of what's likely to happen, in particular where and when.

This detailed knowledge is not even necessary to work the first 200 countries. They come along with only a reasonable effort; in fact, this is one of the problems. Lost in the fast lane of working 100, 150, 175, and 200, the *careful lane* of very rare DX, which may come on once in many years, is often ignored. This is a big mistake. For example, the commitment to search for a 5H (Tanzania), which is not very rare, should be paralleled with the need to keep up with any information about its neighbor, 5X (Uganda), which is rare. Each country is one unit in the count and each deserves special attention. Many current DXers have regretted opportunities lost in their early days.

The hunt for DX is thus a vital part of the sport, and to reemphasize, it is here where abilities vary so widely. The crushing kilowatt and enormous antenna count for a lot. In the finer points, though, experience and knowledge count for more. The situation is not dissimilar from the careful fisherman or game hunter. Long established skills regarding the seasons, the game or fish, the climate on a particular day, and a myriad of other details make the *truly* successful outdoorsman what he is in his sport. DXing is no less demanding, though oldtimers know that there is much laxity between needed operations when the count is high. Young timers know that some of those experienced and well-seasoned DXers are rusty in tough, aggressive pileups. But the realization among the majority of DXers is always there: Careful hunting skills produce rare DX.

One thing that needs to be understand from the outset is that it is hard to be a good DX hunter. I know I'm not the best. One reason is other interests. It takes a lot of discipline to do a lot of tuning, listening, inquiring, and seeking special information. This just cannot be done by everyone. Even with a busy schedule, though, enough background work can be done to make tough DXing easier—much easier. One of the basic concepts is the ability to listen. I know this is repetition, but nothing will be lost more quickly in the mind than the recommendation *not to talk on the rig but to listen to it*. I will discuss some unusual ways to listen for good information.

For those who so receive they also receive rewards: one is pride in finding the DX yourself; the other is a bunch of easy contacts. This work and effort often results in DX contacts that are like gold glistening in a stream. They are sudden and surprising, even if ultimately expected. Of course, some of the intimate DX knowledge is contact with the right people. And from this come contacts with desired stations. This is not widespread, but it does happen. Rarely is it *the* factor which makes or breaks a new country for any given DXer.

I do not reject breaks that come through knowledge of friends. That is, in the true sense, the total scope of DXing. If you know someone who will do you the favor of bringing up rare DX, then you have worked hard for that privilege; however, if you contact someone via a favorable station's phone patch, you have gone over the line. *You* have not worked that station. And how about making the list via the telephone? At first an obvious no. But if you know the list taker, would you call? Tough question. I vote no. I have been in on a few prearranged contacts. No one was bumped or duped in these experiences. There were no telephone calls to break an on-the-air impasse. They resulted from friendships and contacts that were perfectly all right to me, but that others might criticize.

They resulted from a long, patient search of DX matters and did not interfere with other people's contact possibilities. I know of no one who, for any reason, turned down a chance to work a DX station this way.

What is right? I believe a true DXer knows that a legitimate DX QSO originates from his station and was worked by him.

How a given contact is arranged is a source of much debate and few solid answers. In general, the more "remote" the contact becomes, the more guilty the DXer should probably feel. Guilt is ultimately a matter of conscience, though.

The thrust of this chapter will be on searching the bands for DX and searching the bands for DX information. It shall also deal with searching DX friends' minds for information, as well as the many DX bulletins. If this sounds easy, you're in for a big surprise. The delicate ins and outs of DXing require careful judgment and selective aggressiveness, from getting the contact to getting QSL.

To go through the forest with at least some wisdom is better than wandering in the dark. Much about DXing can be frustrating, and any help in the search is useful. Let us concentrate on the hunt, for rare DX is the fox of the airwaves.

GENERALITIES

The ultimate goal of the hunt is to bring about one result: a successful QSO with a DX country. It might be nice if we came into the shack for a round of DXing and our computer printout announced: FR7AI/T-TROMELIN, 21.305, 2230 UTC. BE THERE. Ah-hah, 15 minutes to go. Fire up the rig and get a cold drink. At the appointed time: "K5RSG, this is FR7AI/T. You around?"

"Roger, roger, OM. This is K5RSG. You're five and four, QSL?"

"I QSL. Your report is five and five, QSL?"

"I QSL the five and five. FR7AI/T this is K5RSG. Seventy-three and thanks for the new one."

"Glad to oblige. Au revoir. K5RSG this FR7AI/T now clear."

Ahh, the challenge of DXing! Now, where did I put my drink?

It just doesn't happen that way, though every DXer does have tales of remarkable luck and an easy QSO with a hard one from time to time. Generally, the search is tricky and makes up part of the challenge. And the thrill is palpable, whether you found a new one by simply tuning or by putting together a DX station's pattern and waiting for him on an unexpected frequency. Finally, as one's skills in this area progress, there is development of considerable pride and confidence.

There are many sources of DX knowledge. One may, for example, tune into frequencies where established DXers hang out. Much important information changes hands in these informal gatherings. Another good source for DX information is the nearest DX club repeater. Some hams are too far away, but many and probably most hams can at least hear a DX repeater if an adequate antenna is used. Another source is the DX itself. Let's say you didn't make it one day. Instead of stalking off mad, stick around; occasionally the DX station will announce where and when he will return. Be sure to make friends with DXers *and* DX stations so you can call on them for information. When tuning around, do not look only for DX. If you come across a QSO in which a few knowledgeable-sounding DXers are talking, stop and listen for a moment. Information can sometimes be picked up this way. DX bulletins offer valuable information, usually weekly. Along with actual comings and goings, these sheets also provide tidbits on DX gossip and other DX matters of interest to the fraternity.

What about the controversial elements of the hunt? Is it good DX practice to utilize spotters on DX repeaters? Is a prearranged schedule acceptable? If a buddy of yours is running a list, would you accept preferential treatment? How do you use or limit the telephone in DXing? Questions like these—or the issues behind them—will be covered, but there is no precise moral answer in every case.

Next, the use of this knowledge shall be covered; that is, on-the-air practicalities of transforming knowledge into a QSO. Lists—a big part of DXing—will be covered. Talk about controversy! Just mentioning the word "list" to some DXers can incite violence.

At first it would seem like the hunt should be fun to DXers. In a sense it is, but remember that DXers seem to prefer to be miserable (like duck hunters); actually hunting is a nervous chore. It's not all that much fun when the quarry is not in hand and might not be had after much effort; in fact, one constantly asks himself if he's not wasting his time when long hours on the DX trail have produced nothing.

These silly feelings vanish instantly once the contact has been made. There is vindication and pride for the carefully honed skills. Then the DXer is sure—and he will loudly tell you so—that it's all worth it. Why, he *loves* the time before the QSO and he never was *really* nervous about not getting the contact. He predicted all along that it would be nabbed and, in fact, logged it a few hours sooner than what was first expected. Pure skill, son. It was in the bag from the start.

As mentioned previously, DXers are calm after the battle, not before it, backed up by a challenge to take pulse rates, blood pressures, and brain wave exams during the chase. All true DXers know what the result would show: hyperactivity. This is a nervous group during this particular period. It is no time to discuss with them topics such as crop dusting techniques or how to raise petunias or even national military readiness or stock market trends. After all, *important* doings are under way.

I do want to stress that—while nervous about the contact—knowledgeable DX hunters do have real confidence in their abilities. They know what they're doing and they are completely cognizant that while a given QSO is never guaranteed, they are much more likely to get through than their uninformed competitor. This confidence—in addition to the skill itself—allows for smoother function during the complexities of battle.

CAUTION!

Buyer beware! This book—or any advice for that matter—will not deliver DX into your lap. Personal involvement in DX work is a necessity. It needs to be stressed again. This book is, for example, not a current list of hot frequencies to check out at the moment but a long-term, thoughtful approach to DXing for years to come. The answer to successful DXing lies not so much in being told that the YI is on, which may be true at a given moment, but how to track it down through careful thought.

For this reason, a detailed list of nets and frequencies for particular DX stations will not be found here. The hunt is a *system* of looking around, as opposed to studying the listing of, for example, where the A9 operated last. Of course, such sources as DX bulletins provide just this information. It is important to understand that this *is* vital in day-to-day DX activity. But a book cannot and should not attempt to deliver this type of information. It will be outdated even before publication. Thus, it is important to establish a philosophy of hunting for DX. Nothing will replace your own personal dedication to finding out what is current in DX activity.

If I've hinted about the flux of DX activity, let me *stress* it here. Everything varies. Established nets change times of operating. Random nets change frequency and time. They also disappear when their sole chairmen wear out their egos. DX stations that appear for weeks, months, or even years, may grow weary of the situation and retire quietly. Worse, they may be retired by their governments. Nothing is more unstable than Amateur Radio activity in far too many countries.

Exaggerated? I think not. No one expects XV (Vietnam) or XW (Laos) be on the air soon. But what about other nations who have been on and, for no obvious reason, are suddenly silent? One example of the frailty and unpredictability of ham activity in the world comes to mind: Ethiopia. For months a few spirited amateurs brought this nation to us, and many had contacts with a rare spot. My personal QSO involved a very friendly exchange of stamps and information about our home towns. As of this writing, however, these operators have disappeared from active ham operation. YI (Iraq) is another example. YI1BGD appears reliably on similar frequencies and times for weeks; then it suddenly disappears. Months later, it has reappeared again for a limited period.

The situation is thus very fluid. The DXer, not the DX, is the one who must adapt. As the easy ones are worked, it's tempting to believe the opposite: The DX comes to you. This encourages you to ignore the simultaneous hunt for hard ones which may come and go quickly, to be gone for a very long time.

DX KNOWLEDGE

Ultimately, what we *know* about the DX is what enables us to be effective. This section covers the various sources of DX information and how to utilize these sources. The following section will deal with utilizing this knowledge.

VHF

DX clubs across the world usually are active on VHF. Quite often a club has its own private repeater. These repeaters are for members, but visitors and potential members are generally welcome. My club (Delta DX Association) sponsors a DX repeater. While local DXers outside of our organization may not be encouraged to transmit on the repeater, there is nothing wrong with listening to it. If you are not in a club but are actively DXing, you might want to join a club. True, much of the talk on a DX repeater—on any repeater for that matter—is boring and nonessential, but DX alerts are quite common. When something really good comes up, the repeater is usually very busy with a lot of valuable information being made available.

The type, reliability, and usefulness of this information varies with the occasion and your position in the DX ranks. If you've worked it all and are only waiting for a valid Burma station, you're not likely to benefit much from monitoring. If you are just starting out and several club members are in your position, good information will abound. Many clubs have a concentration of members with fairly high totals, though, and very often the well is dry for beginning information. Beginners don't really need it for the easy ones anyway since, by definition, they are just that: easy. Somewhat rare to rare DX is often what's available.

The information comes in several forms. Very often someone is tuning around, finds something good, and then puts out the DX call and frequency. These are gifts that occur quite frequently in middle-range DXing. *Do not develop the habit of letting this substitute for your own tuning.* It is essential to develop good tuning habits early. *Much* DX will be missed if you rely on repeater information exclusively. Other information frequently discussed on repeaters is related to DXpeditions. The latest news on licensing, travel, expected dates, and such things is often discussed frankly and openly (as opposed to the HF frequencies where these same topics are discussed more cryptically). Finally, when something rare is on, there is frequently spotting for split operation. Spotting is the search for that frequency in the pileup where someone is making contact. This search is usually made by someone who doesn't need the country and is assisting others. Or it is made by someone who got through and announced when he so did.

The first type of repeater information is the announcement that a particular DX station is on and what the frequency being used is. For DXers between 100 and 250 countries, this information can be very useful and provide many opportunities for some new ones. It should always be joined with *your* tuning effort and never relied upon as a means unto itself. For those not in range of repeater reception, this advantage is obviously not available.

Perhaps it is for this reason that use of such repeater help is often criticized. Some people feel that the hunt is—or should be—a personal thing that the DXer should perform himself. Significant DX achievements occurred before repeaters even existed. Their use now is contrary to the sport, particularly since access is not universal. This argument is logical and, and in some aspects correct; however, compared to other controversies that to be discussed, this practice is minor. I suggest we face the fact that repeaters are here to stay. As such, it is just as logical to accept that this is a new—and valid—element of DXing. Though some will lean on it as a crutch, many others both take *and* give. Most hams tune to share the information. In some way, most contribute. I therefore do not see this as a pressing DX issue. DX repeaters offer very good, instant, important DX information for those who want it.

The second type of information on repeaters pertains to DXpeditions. DXpeditions are surrounded by rumors, problems, and all sorts of controversies before they exist. It is difficult for the average DXer to obtain valid information about a given DXpedition. Though there is some secrecy, depending on the information, it is nevertheless true that discussions on repeaters tend to be more open. This is probably created by the feeling that fewer "strangers" are listening; however, in the larger metropolitan areas, there is likely to be more secrecy. Nevertheless, this is still an excellent source for good, hard-to-find news. After all, several members of each club have contacts to find out the right stuff. DXers can't resist talking about all this and for those who listen, there can be a wealth of information.

There is little controversy about this source, because everyone is talking about a given upcoming DXpedition. All are seeking the same information, so few criticize picking up tips anywhere. What's the difference between listening on 20 meters or 2 meters?

The next FM repeater aid is spotting. All admit that this service is very practical. In a rough pileup, if someone spots the frequencies where DXers are getting through, much time is saved and transmitting efficiency is increased. Once the spot is announced, you have the option of tailending there or moving up or down in anticipation of the DX station's next move. In addition, the spotters usually spot the frequency-shifting trend as well, making the whole task even easier. Such activity is usually confined to major DX operations.

Spotting is definitely controversial. Some DXers feel this is spoon feeding. Others say it is just comradeship—they'd help someone when they needed it. In truth, those who criticize this activity will frequently break down and utilize the assistance when they need a new one. Those oldtimers who worked them all before repeaters can afford to give such criticism, of course. What they would do in the current atmosphere is a more difficult question to answer. After observing the situation for some time, I cannot remember any serious DXers who totally shunned such aid. It is certainly not right to criticize people presently doing it if you did it earlier, now claiming it's wrong and you won't do it anymore.

At any rate, it is certainly not illegal. Each DXer has to decide for himself whether it is right or wrong.

It should be clear from all this that—controversies aside—monitoring a DX repeater (or DX informational net on your local repeater) can provide valuable DX information. Clubs are different, of course, and your area may or may not have strong, active membership. The quality and consistency of the information certainly varies according to the club behind the repeater. Finally, if you live outside the range of such a repeater, at least you don't have to face the dilemma of having to accept or reject any controversial DX assistance. Your soul saved, you can now concentrate on *pure* DXing and criticize others who utilize repeaters—until such time as you move closer to the city!

High-Frequency Listening

You should tune the bands not only for DX but also for DX information. It can be found almost anywhere. In general, information can be found in three areas: specific frequencies, usually related to established nets; random frequencies where DXers meet by chance; and finally, informal meeting spots that can change over time.

The various DX nets and how their list operations work are covered in the next section; however, do not forget that frequently the net control or controls discuss DX matters prior to, during, and even after the net. For the most part, these operators are cordial and are open to specific questions. Beginning and intermediate DXers should avail themselves of this knowledge. The level of information usually covers middle-range DXing, primarily the countries run on the particular net; however, these net controls are sometimes in touch with current major DXpeditions and, if information is lacking from other sources, it is often wise to check them for updated information. Remember to do this at the proper time. The proper time is *not* when they are running the DX. It is when the net is open for general discussion.

DX information can also be found on random frequencies on which DXers by chance bumped into each other or perhaps they arranged before hand. It's a bit tricky to find these QSOs, but you don't really have to search specifically for them. What you must do is, while tuning for actual DX stations, be aware of how serious DXers talk. If you simply tune around and reject all stateside chatter, you will miss the subtlety. Here's how to do it.

DXers who really are devoted to this aspect only (and these

are the ones with information) rarely have typical QSOs. They do not get on the air, call CQ, and then chat about the weather, QTH, their jobs, etc. Call this snobbish if you will (remember the beginning of this book), but the truth is they talk about DX. These QSOs are easy to spot it you take a moment to listen. They are obviously different and can be recognized as such. After spotting a few, it becomes second nature. Remember, too, not to waste time with a bunch of beginners or even oldtimers trading "war stories." This is fine if you want to relax; however, serious information will not come from such a QSO. This is not meant to be harsh on general hamming; ragchewing and relaxing have their place. If you're in a hurry and have a specific mission, though, press on and stop on a frequency where it is obvious that DX information is being exchanged.

Remember not to make the search for QSOs among DXers a primary objective. It should simply be a matter of recognizing a QSO in which DX information is being exchanged *while looking for actual DX*. The number of times you will come across pertinent information this way will be surprising.

The last high-frequency source of DX information is the group of unofficial, semi-organized meeting spots in which DXers congregate. Some of these are fleeting. Others are more durable, lasting from months to years. They do change and for interesting reasons.

What happens is that a group of established DXers (who are climbing high) start to hang out at a particular spot. Certain things then fall into place; prominent international DXers stop by; important DX itself stops by; and the DX crowd gets curious and also stops by. Sometimes a list is run, but more often DX information is passed back and forth. As time passes, some of the leaders get bored with the regular appearances on frequency required to maintain control New forces then take over. The frequency of the "regular" meeting place is then likely to change. At first, in a rush of enthusiasm, lists may be run often. Then, as this thrill wears off, the operation may change character and involve what might be called DX forums. Sometimes, good DX shows up and split-off lists are formed. In short, anything is possible, including gleaning good DX information.

For the most part, the discussions in such a grouping are fast paced. At first, the newcomer may think he's not invited. This is only partially true. Most of the DXers on frequency are polite and receptive, and you need only break in and ask. The answers you will get will vary, however. Common DX knowledge—which you still may need—will be generously shared. DX information of a priority nature is more likely to be more closely guarded, depending on how much the people in the know need it for themselves.

For you absolute beginners who hear the big signals of those big DXers and are intimidated, let us set the record straight. You *can* break in and ask your questions. Just do it. Be polite and firm about what you are seeking. Carry through with follow-up questions until you are as informed as can be. You don't have to hang around and be a part of the gang. If you want to, check out when you're finished. Remember that DXers are nice (when not in combat). Your queries will be more welcome than generally recognized.

There are several current DXer frequencies, such as 14,218 kHz. This particular frequency might change with new regulars. If it's not here, it's there. This and other DX accumulation spots can be found easily. Just listen for those serious, dramatic voices discussing priority DX information.

Scheduled Contacts Between DX and Managers

Very often DX operators meet with their QSL managers or other friends at scheduled times. You can check with a given manager at such a meeting and sometimes set up a contact with the DX. Or at least you can learn what frequencies and times are more commonly used by the DX station. If it's not a particularly busy meeting between the two, the manager will not infrequently let you make a contact with the DX station right there. The more important the DX is, the more likely things might get out of hand. Consequently, there are fewer, if any, schedules between rare DX stations and their managers. Or they rotate frequencies to hide more effectively. But for usual DX, which are rare until you have worked it, this can be a good way to find out about a particular station you need or, as pointed out, grab a lucky contact on the spot.

At first you might think that finding the frequencies where such meetings take place is difficult. This is not so. For average DX countries, this information is common knowledge on the bands. If you hear the DX one day but did not get through, listen to his sign-off. He may even announce when and where he meets his manager and invite those who missed him to try on that day. Other sources of these meeting spots are DX bulletins. When a DX station is talking with his manager, someone else will hear it. There is usually a report to the DX bulletins with subsequent wider dissemination. By simply paying attention, you can pick up some of these meeting spots and get some DX contacts out of them.

Listen to the DX

Did you ever not make it onto a list and move on? Tempting. Who wants to sit around and listen to others work a station that's needed? Can't crack a pileup so you shift off frequency? Gets frustrating just to sit there and yell. Well, other things besides a QSO can be gleaned when DX is on the air. If nothing else is active at the moment that you should be chasing, try listening for a while. Very often the DX station announces information about when and where he will appear next. In addition, a net operation might handle other stuff that you need. Frequently, announcements are made as to what they expect to be coming up. QSL information is also frequently available for various DX stations.

The important thing to remember is that much more than QSOs can be obtained on the air. A DX frequency is also usually busy with people who need information. Questions—many of them silly to flatter the DX—are often asked by breakers, some of which may provide knowledge that you need. Don't write off the practice of simply hanging around a DX active frequency and listening.

Friendships

One of the nice aspects of DXing is that special "fraternity" feeling and the resulting friendships that develop among DX stations and DXers. It is nice to get together and swap tales, discuss rigs, and recall complicated antenna installations. Another benefit of such friendships is knowledge. Probably more serious and important DX information is passed through friends than by any other method. The unofficial DX network is vast, rapid, and—considering the possibilities for things to go wrong—is surprisingly accurate. Of course, rumors and falsehoods abound, so be sure of the source of the particular information being sought.

For beginning DXers, do not think that very much information for the first few hundred will come to you this way. Of course, when a rare one is on or coming up, you will benefit from knowledge of others. The point is that you should not expect your old-timer friend down the block to tell where and when a TF (Iceland) or a TR (Gabon) is expected. They are not rare enough to usually make the talk circles among high-level DXers. On the other hand, don't be so busy on the easier ones that you forget to inquire about a TT (Chad) DXpedition. Keep up with both tracks simultaneously.

Another aspect about friendships is that many top DXers become DX themselves, as they activate a country on DXpeditions.

You can't know everybody, of course, but knowing a few and having them familiar with your call sign helps during one of their pileups. This will not make up for a weak signal. Things being equal, however, a DX operator is more prone to pick out a call he recognizes.

DX Bulletins

DX bulletins are an invaluable service, particularly on your way up. I can remember how hard it was to keep up with the tips when

The DX Bulletin

ISSUE 187   APRIL 25, 1983

TWO AMATEURS DEAD IN SPRATLY ATTEMPT; VIET NAM BLAMED

Wednesday, April 20, 1983

THE WORST FEARS of radio amateurs around the world were confirmed today, with official word that two of their fraternity have been killed in the South China Sea. Gero Band, DJ3NG, and Diethelm Mueller, DJ4EI, have perished in an attempt to operate amateur radio from the Spratly Islands. Expedition member Baldur Drobnica, DJ6SI, was reportedly wounded in the attack on their sailing vessell. Others on the "Siddhartha," Norbert Willand, DF6FK, owner Peter Marx, and his wife Jenny Toh, were apparently unharmed.

The unarmed Siddhartha, enroute to an unoccupied reef in the Spratly group, was fired upon by Vietnamese from Amboyna Cay, an island only a few hundred meters in diameter, located near the main Spratly island. The Siddhartha sank almost immediately, with Diethelm Mueller already dead of gunshot wounds. The attack occurred on Sunday, April 10, following which the group drifted for ten days in a dinghy.

Amboyna Cay (April, 1979)

Following three unsuccessful air searches for the dinghy, by the Siddhartha's owners, a Maritime Alert was issued to all vessels in the region. On Tuesday, April 19, the Freighter Linden, out of Japan (Panamanian registry) found the dinghy. It had drifted some 350 kM from the point of attack near

Amboyna Cay. Gero Band, DJ3NG, succumbed on April 17, aboard the dinghy, reportedly of starvation. The Linden was due to land in Hong Kong with the survivors on April 22.

The Associated Press news service carried a story on the situation April 15, but it was picked up by very few newspapers in the U.S. Their sources, Philippine amateur operator Klaus Liller, and German Embassy (Manila) first secretary Jorg Zimmerman, gave the same story as reported in TDXB Issue 186 (April 18). Zimmerman stated that West Germany had "notified all five governments of a radio transmission Sunday (Apr 10) that said the Siddhartha had been attacked and ended with a man's voice yelling 'Fire on board! Fire on board.'" The AP also attributed the stepped-up search for the Siddhartha to distress signals heard on the amateur 20-Meter band on Thursday, April 14.

According to the AP story, Vietnam "first wanted assurances the boat did not carry a political group, and then promised not to interfere with the rescue attempt." There is no indication of why the Vietnamese on Amboyna Cay did not rescue the survivors themselves. Very late word to TDXB indicates that the Vietnamese shelling was extensive, as all six persons aboard the Siddhartha were injured to some extent. They were cast adrift with neither food nor water.

Finally, as for the future DXCC status of Spratly and associated islands in the South China Sea, there has been talk for some time of not counting (for Spratly credit) the associated islands which are controlled by countries other than Viet Nam (which controls Spratly). It is expected that this talk will continue, at a stepped-up pace.

POSTSCRIPT: an AP wire story came in at 1517Z April 20. Please send TDXB a copy if your local newspaper carries it.

Copyright 1983 *The DX Bulletin*

The Spratly tragedy in 1983 was poorly covered in the regular news media. As hams anxiously waited for the news, *The DX Bulletin* dug out the information and provided in-depth coverage for its readers. *TDXB* is noted for its behind-the-scenes news in DXing.

3 September 1984

QRZ DX Hotline (214) 680-1070

Volume XI #10

BOB WINN W5KNE
EDITOR PUBLISHER

JAMES BURTON KB5UT
QSN & QSL EDITOR

JOHN HAWKINS K5NW
CONTEST EDITOR

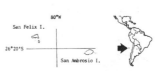

SAN AMBROSIO AND SAN FELIX ISLANDS

CE0AA SAN FELIX
Yes, according to CE3GN, there is definitely going to be an operation from the rare DXCC country of San Felix (commonly referred to as CE0X). This operation, a joint venture by the Chilean Navy and the Radio Club of Chile, is scheduled to begin around September 4-5 and will last for 45 to 60 days.
According to CE3GN, the official callsign for this operation is CE0AA. The callsign CE0AA was selected because it identifies the Radio Club of Chile (CE3AA) operating from a Chilean island.
The operation will be manned by two Chilean operators — Max and Ever. They are both experienced Navy CW operators and both have experience as amateur radio operators (one as a CE9). They may start their operation with some of the DX nets before going out on their own. They will operate both CW and SSB on 80 through 10 meters. Sorry, 160 meters was not mentioned, but listen anyway.
QSL to P. O. Box 700, Santiago, Chile (do not use any other address).
Have patience gang ... the QRM down there will be fierce. Let's not run them off. Good luck.

3V TUNISIA
According to Don Search, W3AZD, DXCC Administrator the documentation for the recent operation by 3V8ZY (IN3RZY) and 3V8AI (IN3XAI) has been accepted they count for DXCC credit. According to Don, most of the operations by stations in Tunisia have been valid operations and are accepted for DXCC credit. One exception, is the operation last year by TS8WCY; it is not valid for DXCC credit.

SV9 CRETE
Mike, SV9AC/SV9, should be active from Crete beginning around September 15. During his last assignment, Mike was OX5AC. QSL via WB4GCP.

FW WALLIS ISLAND
FW8AF meets with the INDEXA Net on 14236 kHz each Friday, at 0400 UTC.

MORE RUMORS....
Ron, ZL1AMO, may be leaving soon for another island DXpedition ... perhaps Pitcairn Island.
A group of North American contesters may activate one of the Chinese (BY) stations this year.

HZ1AB: 160 METERS
[] W7SE who lives in Saudi Arabia, is only a 15 minute drive from HZ1AB. He promises some 160 meter activity on Wednesdays and Thursdays, 2000-2400 UTC. Check 1821, 1828, 1831, 1842 and 1850 kHz. [Thanks DX NEWS SHEET]

The San Ambrosio and San Felix Islands, a territory of Chile, consists of several islands and rocks of volcanic origin, located on the 80th parallel about 540 miles off the west coast of Chile. Over a number of years San Felix has been at the top of the most-wanted list because travel to the islands has been restricted; civilians are not allowed on the island.
In 1981 an "operation" by KF1O/CE0X was proven to be a hoax. The first operation from San Felix and San Ambrosio — CE0XA — took place in 1965 and was manned by group of Americans.

OJ0 MARKET REEF
A operation from Market Reef (OJ0) is rumored for September 3-10. No other particulars are available at this time.

D6 COMOROS
Bill, D68WB, has a schedule each Monday and Friday at 1600 UTC on 14200 kHz with a station in South Africa. He has been known to make contacts on this frequency after his schedule is completed. QSL to BP 540, Moroni, Grand Comore, Republic of Comore VIA France.

POPE'S VISIT TO CANADA
CJ1, CZ2, XJ1-XJ8, XK1-XK2
In addition to the special callsign CZ2P, the Pope's visit to Canada, September 9-20, will be celebrated with several special prefixes. VE1-VE8 stations are authorized to use XJ1-XJ8, VO1 and VO2 stations may use XK1 and XK2 and VY1 Stations may use the prefix CJ1.

ZD7 ST. HELENA
The recent operation by ZD7AL is a bit of a mystery; neither his name or QSL route has been reported. He has been reported on 14049 kHz at 2355 UTC and 21302-21320 kHz around 1900 UTC. ZD7CW has been reported on 21335 kHz around 1600 UTC.

V4A ST. KITTS-NEVIS
The new prefix for St. Kitts-Nevis, V4A, will replace VP2K effective September 19, 1984, the first anniversary of their independence from Great Britain. Non-resident operators are expected to use their home calls followed by /V4A.

Renew early ... don't miss any issues

Hot news on the *QRZ DX* front page. This bulletin frequently provides background information and maps on the more interesting DX countries. It also provides contest coverage and occasional articles on DX tips and technique.

I was in the middle range of DXing. Between my own scouting and the number of possibilities offered by the bulletins, it was a full-time job to coordinate it all. Most of the bulletins come out weekly; thus, the information is usually timely and easily updated. QSL information is also frequently covered for each of the countries discussed.

The bulletins also keep abreast of DX-related stories, gossip, and controversies that make for interesting reading. For those not near a DX club or without DX contacts, the bulletins help keep track of the flow of the sport of DXing as a whole. This aspect of DXing is not covered extensively by the ham magazines, because their monthly formats and advance publishing dates prohibit it.

The basic information covered in bulletins is a country by country listing of anything pertinent that is on, coming on or was on, the last being an update on QSL or other interesting information about a previous operation. *Pertinent country* means any that is usually not readily available. I looked at some typical issues of one of the DX bulletins and noted that the information regarding the following countries was provided: VE/1 (Sable Island), ZL/C (Chatham Island), ZK2 (Niue), 9L (Sierra Leone), S2 (Bangladesh), VR6 (Pitcairn), LU-Z (South Shetlands), 7Q (Malawi), ZD7 (St. Helena), TT (Chad), A2 (Botswana), A6 (United Arab Emirates), TZ (Mali), C9 (Mozambique), and BY (China). This range of DX is fairly typical for the average bulletin.

When something very rare is going to be put on the air, the bulletins usually follow it from the inception of the idea. The last Heard Island DXpedition was followed by the bulletins for over a *year*, as two separate groups charted a long, complicated preparation and transportation course that resulted in both groups being there simultaneously. After the DXpedition was over, the bulletins reported interesting details of the trip. Nowhere else can you find such thorough coverage of important DX. The bulletins are unique in this aspect.

Another useful tool that some bulletins provide is the charting or listing of various DX stations and recent times and frequencies that have been spotted. Using various formats, these are listed simply by call sign with information as to where and when they were found. Using this information week after week, you can track a DX station and often discover an operating pattern. For beginning and intermediate DXers, the information is very useful because it includes some less-rare countries that are normally not covered in the bulletin's primary news. For example, many of the semirare Russians that are not usually discussed elsewhere are listed here.

The way to utilize this information is to chart, week by week, the times and frequencies of a particular station or stations that you need. At first, there probably won't be a definite pattern. As time goes on, there may emerge a tendency to show up on a particular frequency around a certain time. Subpatterns can also be noticed and so identified. While this is going on, you should be tuning as well. The results can be surprising.

I can recall having 9V (Singapore) as a stumbling block. It

THE LONG ISLAND DX BULLETIN
P.O. Box 173, Huntington, N.Y. 11743-0676

THE LONG ISLAND DX BULLETIN IS PUBLISHED BI-WEEKLY AT 100 WILLOW AVENUE, HUNTINGTON,
NEW YORK 11743-0676. SUBSCRIPTION RATES: $12.00 DOMESTIC; OVERSEAS AIR MAIL RATES ON
REQUEST.   DX INFORMATION MAY BE EXCERPTED IF CREDIT IS GIVEN TO THE LI DX BULLETIN.

ISSUE 8 - 84   18 APRIL 1984

N4XX PROPAGATION FORECAST-APR-May

SUN	MON	TUES	WED	THURS	FRI	SAT
			18-HN	19-LN	20-HN	21-LN
22-HN	23-HN	24-LN/BN	25-BN	26-LN	27-HN	28-AN/HN
29-LN	30-BN	1-LN				

Legend: AN,HN,LN,BN=Above, High, Low, Below Normal.
***

CANARY ISLANDS - EA8AAU is active nearly every day on 1841 KHz from 0300-0600 UTC. QSL Mike to his CBA.
***

CORSICA - F5KV/TK, sporting the new TK prefix that replaces FC for Corsica, shows up daily on 14,190 KHz from 2000 UTC. QSL Rudi to the CBA of F5RV.
***

CRETE - 29 April-12 May. DF4RD/SV9 will be active on 10-80, transmitting 22 KHz above the low end on 10, 15 and 20; 5 KHz up from the bottom on 40 and 80. On SSB, check the usual DX splits. QSL via DF2RG.
***

KAMPUCHEA - XU1SS/XU1KC will be active at 0200-0400 UTC and 0700-1400 UTC on Sundays. Frequencies: 14,030, 21,030, 28,030 KHz on CW; 14,195, 21,295, 28,595 KHz on SSB. Low bands by sked only. Also the W7PHO and SEA nets. QSL via JA1HQG.
***

DJIBOUTI - J28DM is active most days on 14,220 KHz around 1900 UTC. Via F6GYU
***

DODECANESE - SV5OX says he is on 1832 KHz every day from 0300-0345 UTC. He then QSYs to the bottom 10 KHz of 80 or 40 meter CW. Stratis is also active on 14,020 KHz about 2030 UTC. QSL to Box 157, Rhodes, Greece.
***

EAST KIRIBATI - T32AB near 3795 KHz daily from 0600-0800 UTC. QSL via N7YL.
***

FAROE IS. - 17-26 April. LA5VAA/OY, LA5DW/OY and OY9A (LA9PCA), on all bands, CW and SSB. QSL via LA5VAA, LA9PCA or the LA buro.
***

FRENCH POLYNESIA - FO8KP likes 7005 KHz around 0600 UTC. QSL via F6GXB.
***

GREENLAND - OX3SG (LA1SEA) surfaces on 21,334 KHz about 2030 UTC and on 14,227 KHz after 1200 UTC. From his new QTH, Helgei also plans a strong effort on 40, 80 and possibly 160 meters. QSL via LA5NM.
***

MALAWI - 7Q7LW is active nearly every day on 3505 KHz between 0330-0430 UTC, then QSYs to around 7005 KHz. You can also find Les most days near 21,292 KHz about 1900-2000 UTC. QSL to his CBA.
***

MOLDAVIA - UO5GQ is near 1850 KHz on CW most weekends at 0300 UTC. UO5OBD likes 14,005 KHz at 1930 UTC. QSL both to Box 88, Moscow.
***

NIGERIA - You can usually find 5N8AFE between 21,300-330 KHz at 2200 UTC, or on 7060 KHz. He's willing to sked you on any of the other bands. QSL to Box 7355, Kano.
***

SPRATLY ISLANDS - 2-8 May. 1S1JZ by DU1JZ, DU3MGL and KE6PU from Panata Cay (the site of the earlier, DXCC-acceptable 1S1CK operation). 10-80 (maybe 160). Frequencies: CW, 30 KHz above the low ends of 10, 15 and 20, 5 KHz up on 40 and 80 meters; SSB in the usual DX splits except on 40 where they will transmit on 7080 KHz and QSX on 7235 KHz. QSL route to be announced.
***

SWAZILAND - 3D6AK works his sunrise peak from 0330-0400 UTC near 1825 KHz on Sundays. He's also active daily near 21,025 KHz about 1730 UTC. QSL via G3WPF.
***

TAIWAN - 28 April-6 May. As part of its celebration of the 5th Anniversary of the DX Family Foundation (Japan), 10 or more JAs have received permission to conduct a "DXpedition." They will bring equipment for as many bands and modes as will be permitted. More later.... 17-24 April by PA0GAM, SM0GMG and OH2BH. 3501, 7001, 14,025, 21,025, 28025 KHz on CW; 3574, 7076, 14,195, 21,295, 28595 KHz on SSB. QSL via OH2BH.

MT. ATHOS - As we go to press, SM0AGD is enroute to the Holy Mountain hoping for permission to operate 11-16 April as /SV0. QSL Erik via SM3CXS.

REP. OF GUINEA - W4LZZ should be opening from 3X about now. Jack expects to be there for 2 weeks and then proceed to D2 and C9. Mostly ssb on the usual DX frequencies and DX Nets.

25 April - Special all-bands commemoration of the 110th Anniversary of the birth of Marconi by IY4MARC from the Marconi home. QSL via I4IKW, P.O. Box 3113, I-40100 Bologna, Italy.

Typical fare from the spring of 1984 served up by the *Long Island DX Bulletin*. The *Long Island* sheet is always crammed with information. Note the variety in this single issue.

118

seemed like everybody else got it with ease, but as luck would have it, this one avoided me for quite some time. It's not *supposed* to be a hard one, so I didn't make any particular effort to get it, other than good tuning when I was looking for other stuff as well. It eventually came to me that a black cloud was over me for this one and I had better do *something* or Singapore would go off the air before I worked it. I started noting the times and frequencies of various 9V stations, and two or three fairly specific patterns emerged. The contact finally came with ease, once I utilized this knowledge.

All in all, DX bulletins provide an excellent source for DX information. If you only need Laos and Vietnam, then such a subscription might not be essential. If you need others and if you want to follow the state of DXing, however, then the bulletins are an excellent source of information.

ARRL Bulletins

ARRL Headquarters station W1AW transmits a special DX bulletin every Friday (UTC). See the W1AW schedule, published in *QST*, for further details. I recall many occasions when W1AW provided last-minute DXpedition information that was available nowhere else. The bulletins also provide DXCC-related announcements.

USING DX KNOWLEDGE

How do you use all this knowledge? The ultimate goal is QSOs. The DX is out there waiting—some of it anyway. For those countries off the air, all we can do is cry. For those hiding, let's go shake the bushes. In your search, remember the influencing factors of the different bands and their different personalities.

The First 100

They're everywhere. Couldn't miss them if you tried.

The Second 100

Most are everywhere too. You could miss a few if you tried, but it would be hard; however, in this group are some semi-toughies. Remember these are not the really rare ones that come on from time to time. These may *numerically* be in your first, second, or third hundred. Regardless, they are special efforts.

Generally, the second hundred is easy; it just takes a little more

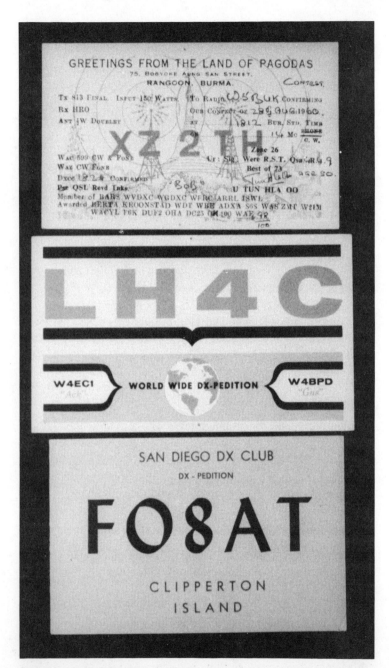

Remember these? Clipperton has been on since, but how many can remember Burma and Bouvet?

time to log them in. Twenty meters in the evening provided dozens of these over and over again. Much is available on 15 and 10, as well.

The thing to do at this level is to *keep hunting*. It's almost tiring to work so much DX and to keep QSLing and QSLing and QSLing. The organizational effort of doing all this, plus applying for DXCC and endorsements is very busy work. Keep at it. Get as many as you can during this period.

Another thing to watch out for is a false sense of security that tough DXing is easy. I can remember going above 100 and thinking it was easy, which it was. Then, during the second 100, I encountered countries I had not thought much about before. These include lands such as 5N (Nigeria), XT (Upper Volta), 8R (Guyana), and C5 (The Gambia). They *seemed* rare to me. Though the names of the countries were known, they were not common. Their prefixes were not heard or talked about often, so their calls *sounded* rare. The pileups were formidable, much tougher than for the first hundred, but there I was getting through. So this is tough DXing! *This* is what it's all about. It's a little rough but I'm handling it! Heck, I'll be on the Honor Roll in a year, two at the most.

Wrong! This stuff was *easy*, my friends. Tough was later—FR/G (Glorioso) on 10 meters, not on in years, as S zero to S one-half on the meter and a huge pileup with other sections of the country more favored. Tough was JX (Jan Mayen), also not on in years at the time, running an S *minus* on 20 meters, a list operation with every other contact nullified because of intermittent QRM and failure to get reports exchanged. Tough was FB8X (Kerguelen) on 10 meters, after making the DK2OC list and waiting for a turn as the FB8 was fading. Tough was Heard Island the first few nights. Tough was FB8WG (Crozet) when first on, as was TN8AJ (Congo), BY at first and XU always.

"Tough" is relative. The countries involved are always not common but can later, after prolonged activation, also not be rare. At a given time, however, their *initial* activation may make them extremely valuable and very difficult to work. Some experienced DXers will recall these examples with sharp feelings.

The point is that certain countries that seem very rare to newcomers are really not. This can induce an attitude that they'll work all of them no matter what. Bad habits can form as a result. For example, DXpeditions—and all the time it takes to track and work them—are ignored. Why concentrate and work one UN1P (Franz Josef Land) when you can work maybe a half dozen other

Russian countries? Why? Franz Josef is not always on like the others. Franz Josef *has to be activated;* though thus far this has been done fairly regularly, the time could come when it will not be so.

So, as you guys and gals take on the second 100, keep things in perspective. Watch carefully for opportunities for rare DX. Always push forward on several fronts.

Semirare and Rare Ones

Sniffing out rare ones sounds like a challenge, which it often is. Yet a little knowledge goes a long way here. There are several aspects to pay attention to in order to anticipate where the DX will be and to aid in the hunt.

DXpeditions frequently operate on what is described as the "usual" frequencies that have been enumerated previously. At the expected time of operation, check these frequencies diligently. As stressed before, do not confine your search to these spots alone. Always check other frequencies as you scan. Quite often there is a QRM on an anticipated spot. The DX station *knows* he will be found and thus is not worried about choosing a frequency away from where he was expected. You have to take this into consideration when looking.

Frequently, the DXpedition team notifies the DX bulletins about the upcoming DXpedition. It is common to announce the frequencies where they anticipate operating. In particular, if they plan to use frequencies other than the usual ones, many DXpedition operators are diligent about telling the DX community. Detailed band-by-band listings of projected operating spots are commonplace. In addition, the more important the DXpedition is, the more hoopla and attention there is surrounding it. In this excited atmosphere, information—and rumors too—abound. It is almost impossible to miss the specifics of the operation, unless you're asleep at the rig.

The advantage of hunting down a DXpedition may be substantial, particularly if rare. When a good one comes on, the very first QSOs are easy but over quickly. The pileup grows like wildfire. Many casual DXers come to think that DXpeditions are on for at least a week and that these pileups will shrink. Though this may occur, it might not. Many DXpeditions are short or are cut short by equipment failure, weather or other circumstances. When the DX is rare and activation is brief, the pileups are likely to be furious for the duration.

Good examples of when this happens are the multiple-stop

Some who were. Trieste, Tangier and the Kuwait/Saudi Neutral Zone—gone with the wind.

DXpeditions. There is usually a flurry of countries, some of which can by be very important. Vince Thompson, K5VT, activated nine countries on one of his sweeps through Africa. One of these was S9 (Sao Tome). S9 is quite rare, and Vince wasn't there long. The pileups were horrendous. Getting him early was a bonus. A survey of needed countries taken before Vince's operation showed that S9 was needed by 51 percent. (This and other references to surveys are from a group of surveys reviewed that were published by *The DX Bulletin*.) This list was limited to DXers needing 100 or fewer DXCC countries. This percentage is thus much higher if all DXers were included. After Vince's operation those still needing S9 totaled 38 percent, a change of only 13 percent points among very experienced DXers. Many people did not get through to this important DXCC country.

Good hunting cannot guarantee a contact; however, those who, by skill or luck, heard Vince when he first came on, were ahead of the game. This is truly an area where skilled searching may reap substantial rewards.

Another little, interesting fact about DXpeditions is the occasional *privy frequency*. The hams who go on the expedition, work an arrangement with their friends in that they secretly listen away from the pileup on a special spot or spots. Those in the know can get easy contacts this way. I've only been privy to the real privy frequency. Perhaps the privy people had all been worked. Perhaps the DXpeditioners had their hands full and were privied out. Who knows? I only know that if *my* friends ever go on a DXpedition to a country I need, I will expect privy contacts. I begrudge no one who has gotten through this way. Could you imagine your best friend going to Laccadive Islands and not working you? Of course not. My wife used to work for an airline. We could consequently fly for almost free. I didn't turn that down, and I'm not about to turn down a QSO with a friend just because he's activated hard-to-get Albania. Thank you.

All in all, DXpeditions are geared for high QSO totals and, as a general rule, most competent DXpeditions usually don't go to all the trouble and expense to put on a rare spot for just a handful of QSOs. The entire motivational force is thus to talk to as many as possible. This is not necessarily true of rare, permanent DX.

Rare and semirare DX operators living in their countries have the DX world by the tail. Example: YI1BGD. Intermittently but continuously on the air for some time, YI1BGD worked a lot of peo-

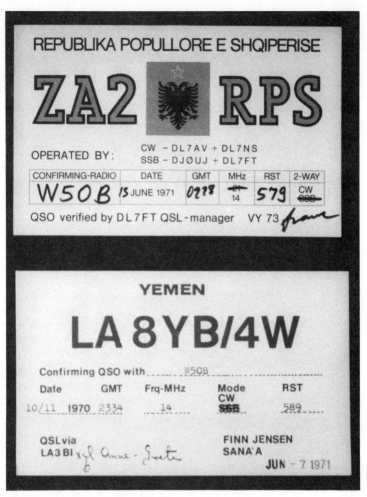

REPUBLIKA POPULLORE E SHQIPERISE

ZA2 RPS

OPERATED BY: CW – DL7AV + DL7NS
SSB – DJØUJ + DL7FT

CONFIRMING-RADIO	DATE	GMT	MHz	RST	2-WAY
W5OB	15 JUNE 1971	0228	~~21~~ 14	579	CW ~~SSB~~

QSO verified by DL7FT QSL-manager   VY 73 *franz*

YEMEN

LA 8 YB/4W

Confirming QSO with .... W5OB ....

Date	GMT	Frq-MHz	Mode CW ~~SSB~~	RST
10/11 1970	2334	14		589

QSL via
LA3BI xyl *Anne-Grete*

FINN JENSEN
SANA'A
JUN – 7 1971

Some we wish were active now.

ple who then poured in QSLs repeatedly for cards that were slow in coming.

The rarer Russian countries provide other examples of hard-to-drag-out locals. Then there's the problem of having very few active hams in some countries. Take Tim Chen, BV2A/B, who until recently was the only operator from Taiwan. Diligent, polite, and a pleasure to QSO, he tries to give Taiwan to everybody he can. He's only one man, though, and he can't work the entire world of DXers. Another loner is Les Sampson, 7Q7LW, the only active amateur in Malawi. Until Les started working DXers more extensively,

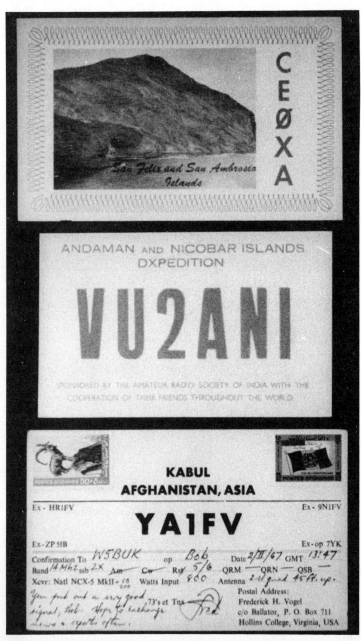

San Felix has been on recently. And hope runs high for the Andamans and Afghanistan.

Malawi was high in the needed polls. Even after several years of increased activity, however, the percentage of DXers still needing him is surprising.

To make this task easier, you need to find out as much as you can about the operating habits of the particular DX station you seek. Occasional patterns of operation make finding and working a DX station much easier. We've touched on this in previous discussions. Let's review some examples. We mentioned before that Pradhan, A51PN (Bhutan), was quite regular on 20 meters in the morning. He could be found on the low end of the sideband portion of the band. Knowing this operating tendency made the hunt much easier; in fact, with this knowledge it seemed more as though he came to me than I to him.

In the early part of China's operation, BY1PK and BY8AA showed up with regularity on specific CW frequencies. Though these shift around, the time and frequency can remain fairly constant for a moderate period-say, several weeks. The new spots or times can easily be found even when there is a shift. Tim, BV2A/B, was a regular on 20 meters, frequently showing up on W7PHO's net. Les, 7Q7LW, had a run on 28,500 to 28,510 kHz from 1600 to 1900 UTC on certain days. Knowing these patterns makes the hunt much easier. In countries such as these where there may be limited activity, it is almost essential to know the operating pattern; otherwise, one could tune for years for a certain station and never find it.

The source for such knowledge is the DX community and DX bulletins. Also, the ham magazines occasionally point out operating patterns. The bulletins do an excellent job of providing current information. After a DXer works one of these countries, he usually ceases to follow said country in detail. Thus, one does not find a lot of talk about them in DX circles; however, someone is always looking, and fortunately the bulletins pick up the information and usually run it.

Hunting for rare ones require persistence and the ability to adapt to what propagation does on a particular band. Remember the Heard Island story. The duration of copyability of this remote spot surprised many. Some things about propagation appeared to be different. During the evening (U.S. time), Heard came on and was audible for hours. This was something of a surprise, because many DXers didn't think that path would remain open that long. When Heard slowly faded and went down, many DXers turned off their rigs. That was unfortunate. Those who persisted and stayed on frequency found that there was a propagation resurgence. Though somewhat of a a surprise, this probably would not have been so if only there had been a station in that area of the world that operated long enough to prove that it could happen. At any rate, it did catch a lot of DXers off guard. When the resurgence came, it was astonishing to see how small the pileup had become. Many were simply not used to this type of fade on 20 meters. To them, the band did sound as if it were closing.

Those who remained got easier contacts. The interesting thing was that this happened almost every night. Still many hams did not learn. Locally, several had to be roused into action. It is of course quite possible that others elsewhere did not observe this effect. It was, to a certain extent, area selective; however, it had to involve a large section of at least the U.S. The point is that, being unexpected, it was not fully utilized. I can remember just sitting there, smiling secretly. As the second peak rose, I snatched a contact on the second or third call. It should be easy to see the main point. Propagation may be unusual. Even experienced hams can be fooled. Astute observation with persistence can thus pay off in DX work.

After all this work in hunting rare and semirare DX, one final thing helps: luck. The constant effort of searching, listening for clues, and calling the DX station finally results in the law of averages bringing you a gift from time to time. If you consider all the things that go wrong in DXing, the random lucky contact is *earned*.

Luck can vary from simply picking a random spot in a wide pileup and getting through on the first call to bumping into a QSO blindly and finding out you've gotten a good one. A friend of mine was in QSO with a stateside station near the frequency China was on one night. The Chinese station picked up my friend's call sign and called him. Those hams monitoring the situation alerted our unsuspecting colleague on 2 meters, who promptly adjusted. He returned a signal report and completed a QSO *he didn't even know had started*. He later got a card for the contact.

I can recall an unusual incident involving a much more common country but a good one all the same. It was 9Q. Zaire. At the time it was one I really needed. Pretty rough pileups accompany stations such as this one because they are not on frequently. Here's what happened. On New Year's Day, I was watching a college bowl game. When the second half started, I walked to the radio room and turned up the rig, set on 15 meters. On the frequency to which I was tuned, there was a list operation running a 9Q station. As

I shortly learned, there was a very large pileup. At that moment, they were on the 5s, and to the astonishment of the net control, there were no takers. I heard him announce:"There must be some 5s around. Any of you 5s need this one?"

Those DX juices flowed and, not knowing what the station was, I quickly popped in the ol' call. He came back and ushered me into the easiest 9Q contact one can have—and with a stateside QSL manager besides. There were still no other 5s. He moved on to other districts where the response was considerably larger.

Why did this happen? The answer is simple. The bowl games on at that time had a concentration of Southern teams. The South is full of football fans. On that day, there was just enough absence on the 15 meters to allow a walk-on contact. It was over very quickly. I was back with my Bloody Mary as I heard the pileup swell from other areas. Shortly after I was enjoying the remainder of halftime with my guests, none of whom were even aware that a DX conquest had been made. My wife might not have condoned this had she known.

Luck in DXing is pure fun. It is like a sudden puff of breeze in a slow sailboat race. I offer these stories as proof that good DX can drop into your lap from time to time. Don't rely on luck as a mainstay, but to a certain extent some luck is a product of hard work.

Established DX Nets

DX nets provide reliable, low-level and occasionally high-level DX opportunities that cannot be ignored. All of these are list operations. All are, therefore, controversial.

To you poor newcomers out there who have to deal with so many lists in the future, I offer you both my congratulations and sympathy. That is the dilemma of lists. They offer much and they take much. They give good DX but they demand fealty. Some are fair and open; others are restrictive. In the end, they are a realistic part of the hunt that cannot be ignored.

The basic goal of a list operation is noble. One station that is copying the DX station well and has organizational ability serves as a link between DXers and said DX during poor band conditions, insufficient language ability, or both. How worthy! Unfortunately, all too often, the translation of this idea into what happens on the air is mutated. The extreme is an omnipotent net control dictating to DX and DXer alike a) who can contact whom and b) who, passing or failing to meet constant criteria, has actually succeeded in contacting whom. It is no wonder that from this soup come the smells of love and hate.

The additional froth that comes to boil and adds to the final pro or con attitude is the belief that lists help lower-powered or "poorer-antennaed" hams. This final no vote comes from those who fundamentally believe that list DXing is not DXing at all but rather a spoonfeeding push of the Pablum into the mouths of DXing babes. All those controversial elements do not even properly align against each other, so a precise debate about the issues is very difficult. Are lists wrong because so many net controls are egomaniacs? Or do they then become all right if they are run by polite, fair controls? Are they fine because they help the little man? Or are they wrong because they ignore basic precepts of DXing? Is a list wrong in a DXer's mind because he didn't get through? Or, after working all the lists that a DXer thinks are possible, is it then right to turn against them? It's like the old unanswerable question: Do you go to school or do you take your own lunch?

You deserve a thoughtful answer, which in the end will not satisfy either side. The answer is: Lists are unavoidable and inevitable. This is not trite thought. The basic spinoff of lists—and what has become of them—is the *production of debates about them*. What is avoided is the very final aspect that lists *are utilized by all DXers*. Why then *talk* about lists? Why boycott them? All true DXers of modern times have been faced with the fact that list operations provide contacts. Why not then yield? That would be too simple. Thus we have to turn and analyze the debate.

In the first place, it is acknowledged that a list operation is acceptable. It cannot be otherwise. It would then be illegal, which it is not. Extending this, the list delivers an opportunity that otherwise would not exist. Thus, in the logic of the situation, acceptance of the list is perfect.

Going against this, unfortunately not directly, is the logic that the list operation is subversive or, at least, unclean in the pure sport of DXing. Let us assume that this is a clear point. We then look to the soul of any and all DXers who have ever participated in a list operation. The question arises: can anyone who has used a list have the right to turn around and damn it?

The issues of disagreement are still present; hence, the war between those for and against. In the spirit of this debate I submit the following concepts. Those lists which are absolutely necessary because of the DX operator's poor signal (for whatever reason), his inability to control a pileup or inability to converse in a language generally understood by the majority of the pileup (this is usually English) are acceptable and even helpful. Those lists, which are run to satisfy the ego of the net control even though the DX station has the ability to work a pileup on his own, are undesirable in varying degrees. Long-established nets that are run fairly have a positive aspect in that they provide a constant, reliable meeting place for DX and DXers. Finally, the list as an entity itself is the result of forces that are irrepressible. As such, it is a permanent part of the DX landscape. Some sort of acceptance of this reality is probably more desirable than prolonged debating, on-the-air fights and the general dissipation of energy that could be generated into active DXing.

Let us now return to the first group of nets to be discussed, the long-established and more-or-less permanent nets. Two very representative members of this group are the Africana and the W7PHO nets, the latter of which has been nicknamed the Family Hour. DX stations check in and the net control or controls take lists. The DX is then run. The process is repeated as long as the DX hangs around or until the net controls end it. (There are other nets that have formal memberships with a long check-in process. Members work any DX they need and frequently the net is opened for others.)

Plenty of intermediate-level DX can be worked on these nets. The trick is to make the list. For DXers with good stations, this is usually no problem. For some reason, list advocates believe that hams with small stations have a better chance at cracking the list as opposed to working the DX directly. I find this unusual as a strong station will be stronger to the net control as well. I believe that in most circumstances, the notion that lists help the little man is a myth. At any rate, cracking a list pileup usually involves the techniques discussed in Chapter 3.

Beginning and intermediate DXers should, if conscience permits (which it usually does), develop a net check habit. Most nets are well run with experienced, fair controllers. The Africana net, in particular, is one of the most well run nets. They are fair *and consistent* in the way they take their lists. This consistency of standards is rare. In addition, the controls are *routinely* polite and helpful, not just now and then. I worked a lot of good intermediate DX in this operation, and I enjoyed it.

The W7PHO net is another well-run list operation. The way things are done does change from time to time, and this lack of consistency makes for some occasional bickering. Bill, W7PHO, himself is notorious for his sense of fair play. Event though he rules with an iron fist, the outcome is usually just. Bill is not often there, however. Yet even so, the net is generally run to his standards. In addition, this list attracts some really good DX. I worked 10 countries through the Family Hour. All were above-average DX countries. You simply can't laugh off this level of DX. The serious DXer must keep in touch with nets like this on the way up.

The DXer should be aware of the many other nets besides these. Their reliability and effectiveness vary. The various Pacific nets, for example, can bring up a lot of those hard-to-get Pacific islands. Here are a few of the nets with frequencies and times which, I stress, can change without notice. (For a comprehensive listing of nets, write to Dieter Konrad, OE2DYL.)

Net	kHz	UTC
Brown Sugar Net	14,310	0330
DK2OC Net	28,750	1100
Pacific DX Net	14,265	0600 (Tuesdays)
PHO	14,250	1500
	14,225	1700 and 2300
	21,335	1800 and 2330
	28,575	2315
Africana Net	21,355	1800
Arabian Knights	14,250	0430 (Fridays)
Seanet	14,320	1200

Random and Spontaneous Lists

In certain situations, unplanned lists crop up. In others, a ham sets himself up as a kind of DX net for different periods of time. Good DX can result from both, and this needs to be understood. Unfortunately, it is in this area that some of the worst things happen, because untrained net controls and egotists take charge and run amuck.

In the first situation, a weak or inexperienced DX station is usually trying to freelance and is having problems maintaining control. Several heroic volunteers scream at him and at each other as they try to put "order" into the operation. Sometimes, someone gains control. If some soul finally does get an agreement to run the DX, what follows is a list operation with a considerable variation in quality. Often, subvolunteers are used to run off frequency and pick up various sections of the U.S. and the world. Meanwhile, breakers, additional volunteers, policemen, advice givers, fortune tellers, angry hams, potential QSL managers, mystics, jammers, hooligans, dissidents, and people with dumb questions proceed to trample the frequency. If the country is not very rare, all this may actually be cleared up. Some sort of list may really be run. Of course, there will still be hootin' and hollerin' on frequency because there is the perception—sometimes very real and sometimes not— that the list was not taken fairly.

The problem here is that the ham "controlling" the net is usually some guy who just wanted to work the DX. In his burning desire, he really *believes* he's going to do a good job of all this; however, after he's gotten his five and five, his interest may wane. Though the list does get run, it may be through pure chaos. Very often the level of DX handled is not worth all of the anger, frustration, thrashing, and gnashing. It happens, though, particularly during the good sunspot years when everybody is a DXer, do-gooder, and citizen of Camelot. And sometimes the job is really done fairly well, particularly if it's a relatively routine country.

If you bump into one of these situations when the country on is one you need, pay close attention to pull out of the confusion those things that are necessary to get you through the mess. Usually, trying to crunch everybody and become the net control is ineffective. There are times to volunteer if a clean, effective opportunity arises. It's a judgment call. Steady observation often results in detecting a sly QSY made by the DX or some other way to get through that is missed by the mob. I once calmly settled a fight between two competing net controls who were attempting to run *simultaneous* lists but who couldn't copy each other. The DX station asked if someone on frequency could copy both stations. I quickly responded that I could. After exchanging reports, I relayed the DX operator's instructions to each of the "net controls." Then I quietly left the frequency with a new one in the log.

One of the unfortunate things about this type of list operation is that often the DX is not rare. You're simply not going to find Macquarie Island being run by just anyone in Frumplberg, Ohio or Bunkie, Louisiana. *Almost* always the DX stations run in these spontaneous lists can easily be found by slightly diligent hunting at other times. Of course, when a country is on that you need at the moment, a confusing list is hardly a deterrent. You'll get in there just like everyone else does. Just play it smart. Listen for the clues and slip in sideways if you can.

The next type of spontaneous list is created by a ham who runs a country or so and then finds himself on a frequency several times a week with some DX check-ins. Usually, one or two fairly good

DX stations will adopt this net control for a while and give him a small reputation and some consistency. This power is enormously inflating to the net control's ego, and these DXers, if male, assume deep and stern announcing procedures. They take and run lists. They announce QSL information. They answer questions. They talk to their friends as if they are big DXers. They handle "rare" DX, such as Hawaii. And finally, *fortunately,* they fizzle out.

Something you need might be run by one of these hams. As with the other type of spontaneous list, however, really rare DX is virtually never run. For the really rare stuff, you must use conventional hunting techniques. Even so, when you find something that you've been on the lookout for on one of these nets, it is best to make the most of the opportunity.

Fortunately, this is relatively easy. For one thing, these nets are not highly visible and are less crowded. Also, the net controls usually know how to run their small show. Compared to the spontaneous list that may crop up with competing controls and other problems, these nets look like well-greased machines. They are easy to break, especially by anyone with experience. These nets are modestly interesting and fairly amusing, more for their pomp than anything else. Established DXers usually pass them by and smile.

Special Lists

Occasionally, rare DX stations which so choose are run by special list operations. These operations are usually quite serious and competitive. Frequently, the people who run them are experienced DXers. Consequently, for the most part they are handled fairly. For a given period of time, these special lists may be almost the only way to work a particular DX station.

When TN8AJ activated the Congo, it was in high demand. The operator was an East German ham. Several Midwest and West Coast DXers ran a highly organized list for him almost every Wednesday. FB8WG made long-needed Crozet available. He was run on some established nets, but also by a special list set up just for him alone. Both of these operators eventually went on to some freelancing. With occasional checking in to nets, they became easier to get. At first, however, if you wanted to work them, you had to go through the particular list operation.

Because you rarely know if and for how long a given operation will continue and because countries like these are too important to be ignored, you simply have to get involved with these special lists from time to time. There are several things for the DXer to know to help crack these difficult operations.

It is mandatory to arrive on frequency early. Important announcements regarding how the list will be taken are made then. Usually, one or more stations will slide off to other frequencies to take the various areas. You must pay attention and find out on what frequency your call area will be taken. Also, pay very close attention to find out which of the control stations will be taking your area and where he is located so you can properly orient your antenna. Then listen for the actual announcement as a particular segment of the list taking is about to begin. Invariably, various hams flub this again and again. You'll always hear desperate voices querying as they try to learn what is going on in the list.

Once the list taking starts, the usual pileup techniques apply. The entire thing moves pretty fast because limited numbers are picked up.

If you make the list, only the wait remains (and the nail biting). The technique part is done. Now, the DX station will show, he'll be strong enough, the jammers and other QRM will be minimal, and the DX will not leave before your turn—you hope! Lots of things can still go wrong, and it's not at all unusual to have to try several lists to get through.

One problem that can crop up with any list operation is that the list taker may be in unfavorable location relative to your QTH. This is particularly true if the distance is small so that signal levels are markedly reduced. In fairness, good net controls designate scattered list takers. Areas can be further subdivided (by state, for example), which eases the burden. As one goes through the "list experience," this can be a factor that sometimes helps when in the propagation sunshine and hurts when "in the shade." It can be very annoying. Once again, a strong station is better than a weak one in trying to make up for any propagation shortcomings.

Other than debates on antennas, I can't think of anything that can get DXers into more pivotal points of view. The value of nets and their negative points have, I think, been fairly presented here and in the Amateur Radio media. The concept of having someone "carry" you to a DX contact is less than optimally desirable; however, in the nature of sports, very few challenges are totally individual. Many are team efforts one way or another. Even chess, tennis, and boxing champions work on theory, tactics, and other subtleties with varying forms of help, advice, and sparring. There can be no final editorial judgment here. These considerations are personal; however, beware of judgments made by others for you.

Any DXer above 225 or 250 countries has most likely stopped by a list watering hole more than just a few times.

Prearranged Contacts

This was briefly discussed in the opening sections of this chapter. There is not much to relate here about actual practice since this involves the cultivation of friendships and personal contacts and usually such an opportunity arises without solicitation. I separate this from the next section in which you contact the DX or his QSL manager. The prearranged contact which comes to you is almost always a gesture of someone who wants to help you for any number of reasons. Perhaps the other station is genuinely trying to assist you in getting a needed contact (most often the case) or perhaps will claim a similar favor from you later. DXers always help each other. Whatever the reasons for the opportunity, it is simply up to you to decide whether or not to accept and be on frequency at the appointed hour. These occasions are not common but do crop up from time to time if you are active in the sport. Simply be aware of such opportunities and be prepared to respond to same.

Direct Contacts with DX and Managers

Another way of getting DX QSOs is to write to the DX station or his QSL manager directly and try to set up a schedule. This is not infrequently done with China, and the BY often shows up on time at the appointed frequency. An easy contact can be had. When setting up such a contact, use alternative days as well as additional choices of frequencies and times, depending on communications reliability to the DX locale. The manager may send information to you about the usual times and frequencies that his DX associate operates on, or he may invite you to join him on one of his skeds with the DX station and put you through.

I certainly do not recommend using this as a primary DX technique. Writing dozens of letters to DX and managers across the world will not result in high QSO rates. Standard hunting techniques will be far more productive; however, if one particular station is eluding you, it is often helpful to write and at least get information about the DX station's operating patterns. And on that unique occasion when a scheduled QSO results, take it for the long-shot good-luck chance that it is.

WFWL

There is a DX saying in those letters: *work first, worry later.*

Lots of things come on the air, including *pirates* (illegal stations) and other assorted "countries" and islands. Many times these stations are not what they represent themselves to be. Frequently, the attempts are genuine even if the results are not. For example, many sincere efforts at activating certain countries take place that ultimately do not pass DXCC scrutiny because of the lack of valid documentation.

The thing to do is work 'em. After the QSO, you can then worry freely about whether or not it was a country or, if it was a recognized territory, whether or not the operators were really there and if the proper license had been issued and so on. You can worry about the logs getting back safely and whether or not you're in the log. You can go further and worry about your QSL being filled out incorrectly. There is no end to the things you can worry about but if you don't work them first you may have something to *really* worry about. Consider the following.

When the Sovereign Military Order of Malta (1A∅) first came on, many thought that the territory (inside Italy) would *never* count as a DXCC country (which it eventually did). Those careless, then, about working the station now have a substantial problem. 1A∅KM comes on infrequently. *That's* something to worry about, and apparently the country will remain rare because of restricted operating.

A similar situation occurred for A6, United Arab Emirates, though for different reasons. Amateur Radio was banned there in 1979, but a few stations operated over the subsequent years, including A6XJC. Because the A6s on did not count for DXCC, many DXers ignored XJC. The time for relatively easy contacts continued to pass by. Acceptable paperwork for A6XJC was finally obtained, and those who had made the effort suddenly had a valid A6 card. Later it was reported that the operator left the Emirates; once again A6 became more difficult to work. *WFWL* is therefore a most valid concept. Foreign Amateur Radio operations are just fluid enough to offer the occasional DX surprise. Each opportunity must be taken somewhat seriously even if doubted at the time of operation.

Fair Play

Americans use the telephone a lot; we live by it in our work and personal communication. It is a relatively benign, respected instrument. But tell a couple thousand hams struggling to get on a list that a given station just made the list by calling the list taker on the phone and you'll have some fire on the frequency. In DXing, the telephone is just plain controversial.

The basic concept of DXing, particularly in relation to the actual on-the-air working of DX, is, or should be, the final, personal challenge which you, yourself, overcome. Thus, if somebody in a favorable QTH is patching your voice and call and you contact the DX this way, you have an obviously unacceptable QSO. But what if you're on the phone with someone who is simply spotting for you? Or while you're transmitting, the guy on the phone says to stop calling, that you've just been answered by the DX? If you take this assistance, is your QSO valid? In many respects this is identical to the help commonly found on repeaters.

The big question of whether or not it is fair play to get on a list by phone is also not easy to answer. It sure *sounds* wrong, but some would argue that if you developed the friendship or contacts to do this, you have earned the right to exploit it. The problem here is that the list gets stacked in advance and some on-the-air stations are not going to get on it as a result. Because the list operation is being touted as fair by the net control, it is, in fact, not an even-handed net tactic.

Though most of these issues are not black or white, I do feel the following general principle applies. The more outside insertions there are that form or penetrate the events of your DXing the more the degree of remoteness between you and your DX contact becomes. How much remoteness you allow is an individual response and is to some extent based on your feeling toward or your value of the various DX awards. Most DXers will not turn down reasonable help in the thick of battle; however, most know when the line is crossed and the spirit of the sport is breached. Outright cheaters exist in every sport. In DXing these people know themselves. They also know the value of the DX awards they hold.

6

HELP!

There are devices, objects, circuits, and other such things that help the DXer. This is a "non-skill" area, but one which is very important all the same. It includes everything from equipment that is not part of the primary station (such as speech processors) to such miscellaneous stuff as callbooks and antenna bearing charts. It also includes the physical presence of awards; i.e., they must be on the wall. (This is a vital point of DXing.) And, finally, DX clubs, those bastions of confident ladies and gentlemen who know all there is to know about DX. These are selective fraternities, and here their innermost secrets will be revealed.

Before covering the different aids separately, one area of special interest should be reviewed. You probably have local competition that you would like to beat out in DXing. The reason may be that you must want to be ahead of this ham in your DXCC count. Or he may live close to you and offer direct on-the-air competition for new ones in addition to providing heavy QRM when he is transmitting. The later is a serious problem as it is difficult to copy DX when a nearby ham is calling on or close to the frequency. It is therefore essential to eliminate this competition *before* it gets on the air. And you can do it with a shack-full of accessories.

The thing to do is stuff your radio room with vital magic boxes. The operating desk in particular must be excessively cluttered. One really beneficial thing about clutter is confusion, in particular confusion of the *enemy,* your ham neighbor, or friend who is your competition. To do this right takes thought and care. Yet it can be devastating to your fellow DXer.

Get as many boxes, cables, blinking lights, dials, charts and related things as you can find. Knobby things are especially important as nothing impresses a DXer more than knobs. Next, interconnect all this stuff in the most confusing way possible and be sure to almost bury the rig in it. On the main and side desks have stacks of DX charts, beam headings and lists which are marked *secret* in red ink at the top. Finally, record a tape in which you camouflage your voice to sound like rare DX. Keep your receiver on and close to the recording mike for the authentic band sound. Call CQ and sign something really good like 3Y (Bouvet). At the beginning, record lots of QRM, making the signal almost unintelligible but gradually clear it up for reasons that will shortly become obvious. After several CQs, record a QSO with the DX calling you. Be sure to leave breaks in the tape to allow you to transmit. Connect the tape recorder up so you can trigger the playback from one of the many knobs at your command.

Now, you're ready! First, drop a hint to your friend on the air or by phone if necessary that you've updated your station for *serious* DX work. Don't say anything more! Don't answer any questions. Be evasive and vague. As your friend's nervous factor rises, he will start to hint for an invitation to come and see the new stuff. Resist at first but finally give in and allow him to come over. Make sure that this is at night so you can lower the lights in the shack which makes all the equipment much more mysterious in appearance.

When he comes over usher him into the shack. Put a sign on the door which says something like *DX Central* and be sure to have the door locked so you can unlock it in front of him. This is a very impressive maneuver. Once in the room you start to turn all the stuff on, rapidly of course so as to add authority and confusion. As you're tuning around, your visitor is sure to notice your DX charts and tables and he will at some point spot the papers marked secret. If he tries to pick one up, immediately stop what you're doing and dramatically lurch toward the papers while simultaneously screaming "No!" as loudly as possible. Give him a fuzzy explanation about these being available from special DX information sources which you promised not to reveal. Next, hunch over the documents so as to deliberately conceal them and flip through the pages loudly. Then, turn to your friend and say, "Bouvet should be on 20 meters any minute now."

Next, tune around a bit and announce that you are going to put in the DX signal injector and short path propagation enhancer. Turn a few knobs, one of them being activation of the tape recorder and quietly turn your receiver gain down. As the CQ comes on, tell your friend that that should be Bouvet. Of course there's all the prerecorded QRM. Start twisting and dialing more knobs while informing your friend that you are activating the remote interference slicer and DX readability phase protector. As the signal begins to clear up turn to him and say with a bland expression, "Remarkable, isn't it?" Next, announce that you are going to employ the external QRM defractor array and the synthesized half-loop full-wave noise splitter. At this point the signal is isolated and clear. You now proceed to "work" the station, as you had set it up previously on the recording. The pauses allow you to really transmit which should cause some interesting responses on the air. These can't be heard in your shack, of course, since the receiver is quiet. Finally, conclude the QSO and turn the equipment off.

As you turn to your friend, you will notice him to be a pool of sweating jelly. You now quickly explain that you're not sure if this one will count as the licensing is uncertain. This covers you for not submitting the card for this "QSO" and adds to the confusion if your friend tries to track down the operation. In addition, it sounds like you sure know a lot of things. While you are talking, turn the papers marked secret over so as to give your friend a little chest pain.

Of course, the questions come. A preset alarm should go off reminding you that you have a schedule with King Hussein, JY1, and that your visitor has to go. You tell him that you'll find him in shortly and usher him out before he has a chance to field any questions. Then, avoid all contact with him for two months. This will add to his confusion, frustration and will greatly increase his lack of confidence.

And now—the results. One or more of the following is likely to happen:

• He will waste a lot of time searching and researching the whereabouts of all this stuff. You'll be on the air.
• He will spend a lot of money trying to buy everything which will reduce the amount available for upgrading other aspects of the station.
• He may become so disgusted with all the junk that he'll go through a period of markedly reduced operating time.
• His or her spouse will become angry and possibly restrict on-

the-air activity. Even better, there may be a divorce requiring the ham to sell the rig to handle the settlement.

As you can see, there are many possibilities for reducing or eliminating local QRM. After all, DXing is serious. Serious measures should be taken.

On with the program. The following helpers provide varying degrees of assistance and some really are important.

SPEECH PROCESSORS

In my opinion, speech processing is absolutely essential for SSB DXing. Period. Fortunately, processing is becoming accepted as a standard of operating whereas it used to be criticized.

In DXing, one must employ some type of speech processing because each pileup is composed mostly of processed signals. Processors are now common components in currently manufactured equipment, and thus even casual DXers have improved and competitive signals. In fact, in many of today's rigs, processing is standard; thus, it's difficult to avoid having it. Then, one must only get used to the idea of turning it on. I should add, however, that other DXers feel that a loud, clean signal, without processing, stands out in a pileup.

If DXing requires a processor, and almost everybody has one, then what is to be done for a competitive edge? Answer: *two* processors. Many will now recoil in horror. The loudness! The distortion! It doesn't even sound human! Well, folks, it can be clean *and* it works. In fact, it is with some reluctance that I relinquish the information here as it really is effective and not many people understand how to set it up. Unless properly done, the results can be disastrous and, in the end, counterproductive. Even so, many have dual processing with excellent signal characteristics. Those who take the trouble to set this up correctly spend the time to make sure that the final signal is worthy of being on the air.

Here is what I suggest. If one is to utilize two stages of processing, one of the levels should be internal. Though undoubtedly some people have worked out successful formulas for more than one outboard processing system, the work to keep this clean involves more than what is recommended for the average DXer. Internal processing usually has the bugs and kinks worked out. The same is true for accessory processors which are not originally included in the rig but are carefully designed for installment with due attention paid to problems of distortion and feedback. It is not particularly essential that the internal processing be audio or RF, and I will not attempt to enlarge here on the theories of which is best. Suffice it to say that almost all modern processing systems are effective.

When selecting the outboard processor, try several units owned by friends to see how the combination works at your QTH. Test not for effectiveness (that is automatic with good processors); instead, focus on cleanliness and clarity. Dual processing has a "heavy" sound and it is easy to confuse distortion with punch. I urge anyone who tries this to spend as much time as necessary to work out the details.

There are some who would say that this is foolish. RF is RF and, to a point,processing enhances the signal and then does nothing more or perhaps even harms it. True. However, dual processing—when done right!—edges your signal above the others. Everybody is processed. You might as well be processed better. This is one of the realities of DXing.

I will now relate my personal experience in order to prove a very valid point—that amplifiers and antennas do not have to be the best. When I made the transition from being a casual DXer to a committed DXer, my station, in its entirety, could not suddenly change. And, truthfully, it did not need to change immediately. After being on the air for twenty years, I still needed Paraguay, for example, so I did not require high tech gear to make such contacts; however, as time evolved the pileups got tougher (imagine that!), and I figured out that some sort of transmit assistance was in order. What happened, sequentially over several months, was that I discovered the advantage of multiple stages of processing. First, I purchased an outboard processor and later an inboard one. My effectiveness changed dramatically, and this is with the use of a 1200-watt linear and a two-element quad. Heavy station, huh? Hardly. And yet I reached a 293 country total.

What happened? Was it my "powerful" amp? Cough, cough, laugh, gasp. Was it my almighty two-element quad still turned by a TV rotator? Heh, heh, ho, ho. Yawn. Was it my *skill*? (Yes, yes, yes!) Or was it those processors that pumped my little linear to its limit? Well, the truth is that since I installed the device myself I can attest that I suddenly became a "big" signal. It took time to iron out the bugs but that little station could work anything.

When setting up such a system, it is essential to perform as much testing as necessary to ensure that the signal is clean on all bands, with the antenna pointing in any direction. This last point is very important. Frequently there is a direction (or there are several directions) in which RF is radiated more heavily toward the radio room causing feedback problems. This is due to varying factors and must be tested at each QTH. Failure to do so can result in unexpected distortion and signal deterioration in certain beam headings. You cannot assume the signal is clean if you run your tests with the antenna pointed only one way. In general, connecting cables of processors should be as short as possible and shielded. Power supplies and their connections should be checked as possible sources if RF feedback is present. RF chokes can be installed in difficult situations. With adequate testing an effective, clean dual processed signal is possible.

The results are impressive. It has been assumed by most hams worked that my station was bigger than it actually was. Face it, 1200 watts *input* and a two element quad (which I still use) are not big gun material. And though I wasn't always first with the contact there were a lot of toughies in that 293 total. Recordings of the signal by distant hams (Germany and Australia) in which the processing was brought in stepwise demonstrated the effectiveness of the system.

One final note: Remember to turn most of this stuff off when you're not DXing. Dual processing is for pileups, not ragchewing. (Please note another approach used by DXers—equalizing audio response for your mic and rig with an equalizer.)

ACTIVE AUDIO FILTERS

Audio filters have been around for a long time but the present generation is pretty sophisticated. With receivers at the level they are now, however, these filters are usually not the critical factor between making or not making a contact. Nevertheless, they can be beneficial. If your rig needs more selectivity, its performance can be very helpful. I remember when I put a 500-Hz audio filter in line with my transceiver, which had no alternative selectivity other than the basic 2.1-kHz mechanical filter. On CW it was like a miracle.

These filters tighten up audio selectivity for sideband and CW with high-pass and low-pass filtering combinations. The limitation is that interfering signals in the passband still have some blocking effect and the desired signal is not selected out as well. The circuits have gotten fairly sophisticated, however, and all in all they do a good job. Some units have circuitry which electrically isolates desired CW signals and produces a note which is almost noise free. These filters can be very narrow and when following a sharp IF filter can really tighten up the whole picture; however, in the presence of very sharp CW selectivity, this final tightening is not really essential.

Most of these filters also provide a notch. Notch circuitry is becoming more standard in today's rigs but some do not include it. A notch filter is very important in DXing. I recommend an outboard audio filter for this reason alone if one is not present in the receiver or transceiver. These filters usually provide good notch depth of up to 60 or 70 dB. There will be more discussion on the virtues of AF versus IF notches in the chapter on station equipment.

I haven't tried all the various brands, but people on the air have generally been pleased with whichever one they have chosen. There are also no models that I have heard of that have drawn universal complaints. All in all the various units appear to be capable. There are some rigs with matching speakers that contain audio filters. These are not nearly as fancy as the individual units which are more competively manufactured and I would not consider these speaker add-on's to be front line active audio filters which are much more sophisticated and effective.

KEYERS

Keyers are essential in DXing. Bugs have their nostalgic use too and the fun of operating both straight and bug keys will, I hope, never die out. However, in DXing the clarity and correctness of your call while being sent are extremely important. Keyers provide the best dot-to-dash ratio so that you can weigh dot length and dash length precisely as you want it. All this can greatly enhance readability at higher speeds. What's more, a properly weighted call stands out against others on CW and makes you more prominent in a pileup.

Most keyers provide some degree of memory capability. I strongly recommend selection of a keyer which has this function. In a rough, prolonged pileup, it is easy to become tense and occasionally send your call incorrectly. It is enough of a job to clear up a miscopy of your call in a busy pileup when it is sent correctly. It's even more difficult if the DX station copied exactly what you sent, but you sent part of it wrong! Use of memory to ensure that your call is sent correctly every time makes for more reliable, trouble-free contacts.

There are many models to choose from, and it's easy to get opinions and the chance to try several through friends. I have no strong recommendations here but features do vary and you should carefully weigh your keyer needs before purchasing. Paddles for keyers come in a good selection. One important thing to watch is the base. The paddle base should be heavy enough so it doesn't slide around when you're trying to operate.

CALLBOOKS

When you first start out, *Callbooks* are essential. There is a ton of QSL work that has to be done and *Callbooks* are absolutely necessary. I—and my friends—remember this well. I got bitten by the DX bug as the year 1979 was well under way. The new *Callbooks* were coming out in a few months and I didn't want to order the older set that was ready to be outdated. So for those months I leaned on three of my friends continuously. You could hear the moan when they answered the phone and realized it was me, then replying before I even said anything, "Hold on, I'll get the *Callbooks*." In the background I could hear cursing and grumbling which was very disconcerting. After numerous insults and threatening to give me wrong addresses they would finally give some QTH info. Then, in dire tones, I would be warned that that was my limit and not to call back for the rest of the week. One of the hams involved, an XYL, did most of this work for me and now I suppose she'll want a free copy of this book. We'll see.

At any rate, callbooks are essential. The primary publisher puts out volumes titled *Radio Amateur Callbook*. It comes in two editions, Foreign and U.S. Both editions are needed as many DX stations have stateside managers. In the next chapter I will discuss the in's and out's of QSLing and how and when to use the bureau or go direct, or both. Suffice it to say that for serious DXing there will be much direct QSLing and, in the case of managers—foreign or stateside—one needs specific address information.

The *Radio Amateur Callbook* is published annually. Besides its ham listings, other DX aids are included, such as a postal rates chart, addresses of various bureaus by country, and great circle bearings.

DX COUNTRIES LIST, MAPS, BEAM HEADINGS

Obviously, to work DX one needs a countries list to identify a prefix heard with the country to which it belongs. When starting out, the less common prefixes and country names are usually not on the tip of a your tongue. With time many become second nature but initially UK8J vs. UK8H, UK2B vs. UK2G and TJ, TL, TN, TR, TT, TU, TY, and TZ can be confusing. The easiest way to get an accurate list is to simply send for the *ARRL DXCC Countries List*. (If you don't send the dollar, you'll wait even longer for the list. Actually, you won't get it at all!) I put mine in one of those plastic protectors and keep it at the operating position. Because countries use unusual prefixes from time to time, you should also keep available a supplemental list of international callsign series (found in the *ARRL Logbook* or *ARRL Operating Manual*). Thus, a mysterious RW2 station can actually be traced to White Russia, not nearly as exotic as the prefix at first suggests.

In terms of maps, the brand new *ARRL Amateur Radio Map of the World* is superb. For U.S. hams, the center of this world map is the States and a world's eye view of how the signal travels is dramatic. There is distortion in the land masses portrayed, but this is actually a correction of how a radio signal "sees" a given target. Australia is therefore spread wide in an up and down direction on this map, but you can really appreciate how to point your antenna for the different areas of this nation. In fact, with the exception of the central countries, everything is stretched and distorted from a conventional map's point of view but it is amazingly accurate for what you are trying to visualize: your signal's path across or around a sphere to its DX target. I obtained this map early in my DXing career, and I turned to it then as I turn to it now for an appreciation of exactly what I'm doing with a long-distance contact. It is truly one of the most creative, unique, and effective maps I've come across. And this is from a sailor who likes "flat" maps and straight projections.

Further, the map shows the different time zones. Of course, it also shows the prefixes of as many nations as can be printed or fitted on to their surfaces. And finally, it enumerates in the margins that one thing you occasionally require, an alphabetical listing of the countries by name with their usual prefixes following. This is not needed often but when someone alerts the Rwanda is on, it's nice to be able to reference the list on the map and determine that the prefix is 9X. This touch complements the ARRL DXCC List, which of course goes by prefix alphabetically, and allows for rapid cross referencing. I enthusiastically recommend getting this map on your wall.

A list of beam headings is also very important. My DX club issues a list of all DXCC countries in very small print that is one side of standard typewriter size sheet. There are four large columns to handle the considerable number of countries with very small columns arranged as follows. The first column has the prefix. In the case of countries which have and use more than one prefix (such as the Russians for example) the most common or first prefix is selected. Next to the prefix is the country's name and the column next to this is both the shortpath and longpath beam headings for that country.

Our club also puts a computerized list of all the members and what each member has confirmed. This is primarily used for an alert system if a country comes on that some members need. This too contains one's own column of confirmed and is easily updated but with all the members' listings it runs to ten pages (though this does include other DX data as well). Thus, even with computerized assistance, that original single sheet with all the countries on one side is still the most useful and fast way to get a beam heading or determination if something I hear on the air is needed. Whatever you arrange, have some easy system for getting that beam heading quickly and an easy way to see if you have worked and/or confirmed any given country.

There are plenty of commercially manufactured lists and aids which run from helping you organize contacts to following the gray line for DX communications. You don't need—or even want—all of them, but you need a system for tracking the factors I've covered in this section.

EARPHONES

There are times when signals fade that earphones can make the difference in copying or not copying. One should have a set of communication type earphones in the shack for those times. Do not use stereo or wide frequency range type headsets because they enhance noise and other sounds along with the signal. Communications type units are better because they restrict the frequency response and promote enhancement of certain voice characteristics and the CW beat note.

I found a pair of headphones designed for use in tanks. (That's as in *military* tanks.) As you can imagine, these vehicles are quite noisy, and reliable communications among crew members within the tank and externally by radio is essential. These headphones are designed to enhance speech, and they have an internal acoustic ear piece which fits directly into the ear in addition to the regular rubber cushion that fits around the ear. The beat note of CW is enhanced nicely as well. I bought these at a hamfest for a few dollars and I would imagine they would be available at some Army surplus stores. If you buy a pair, be sure to note that the left and right earphones are separated (like in stereo) to allow different communication channels within a tank to be active at the same time. The plug wiring can be changed to make them monaural.

MICROPHONES

Microphone selection is an important aspect of DXing, but it is becoming less of a DXer's perogative to choose because of certain manufacturing tendencies. Not too long ago, rigs were made using two types of mic jacks which allowed you to select which microphone you wanted. The use of varying DIN jacks now increases the complexity of what will plug into the front panel. In addition, many rigs (predominantly Japanese) have particular matching circuits for microphones that allow only their special microphones to work well with the rig. They then sell their mics, and the tendency to buy it and not modify the rig for a preferred mic is strong.

One argument for this concept is that the total audio input circuit is balanced and matched for optimum performance. This is true but may not be true for optimum DXing. Most of the microphones used are dynamic and, as a result, have more bass. Combined with built-in processors there is considerable transmit effectiveness. DXing needs are different, however, and certain microphone and processor configurations can improve on these rigs. The problem is that a manufacturer looks for a middle of the road solution: a combination of dynamic mic and processing that is not too shrill and has more of a broadcast sound. Fine. But DXing is not a middle of the road sport, and *when* DXers have a choice many will play around with mic selection. Many prefer higher pitched, more piercing audio that can be had if they get to put the mic in that they want.

Some will modify their rigs to accommodate their choice of microphone. But since this tampering may cause warranty problems and possibly affect resale value, most just accept the situation and, with time, forget about the nuances of microphone effectiveness. The difference came home to me when I transmitted on a typical dynamic mic (Japanese-rig configuration), and the locals (who did not know I wasn't on my regular transceiver) commented that my audio was flatter, had more bass, and lacked the punch that they were used to. Everybody said it sounded "good"— there was no fault in the audio—but something was different, and most said it was not for the best. Again, that nice, middle ground, pleasing audio is fine; however, if the option is there, plan your mic needs around DXing. It should also be emphasized that a lot of the decision is based on personal preference, and it's difficult to get people to agree on audio reports. Still, I do recommend playing around with different mics to see if some type of consensus on your sound can be formulated.

I have certainly not experimented with every microphone available, but I have made the effort to run through some. In making my selection, I relied more on what people told me, rather than judging recordings of what I sounded like to me. And in gathering experiences from others, a crystal mic seems to emerge as a good choice for DXing. This is not to say there aren't other fine mics; of course, there are. Yet, every time I run tests, almost everyone— even those who have chosen other mics—tells me that my crystal mic sounds best for its cutting edge.

This is only one example of a preferred microphone. Many DXers have found other mics that accomplish what they want for their voices. My main point is that the experimentation and selection possibilities are gradually being diminished by the manufacturing tendency to mate mics and rigs in special circuit configurations. If you have a rig that allows you to choose, do so. And if this is a subject you haven't paid close attention to and you're using some random mic, then get on the air with some alternative choices. You'll be surprised at the differences that you and others will hear.

MICROPROCESSOR CONTROLLERS

In Chapter 9, microprocessors will be discussed in general; however, some rigs offer touch pad options for controlling the microprocessor and I will cover this here. With this accessory, frequencies may be dialed up quite easily. In addition, storage of many frequencies is possible which makes for easy tracking of several DX stations or several likely frequencies when waiting for a DXpedition to fire up. Depending on the arrangement, the frequencies stored may be run through or scanned rapidly by pressing a single button. And, needless to say, if you're on 28,500 kHz and someone puts out a repeater alert for China on 14,049, you can QSY very quickly.

Interestingly, many hams are not enthusiastic about this option. Of course, it's not available on every rig. However, when it is, I endorse it. It does take using it to fully appreciate the flexibility and speed it provides. Afterwards, it is something you won't want to part with. I predict that a lot of manufacturers will be employing this feature. For DXers, it really provides assistance.

AWARDS

There is an upcoming chapter on the DXCC program and other DX awards, which is the central core that gives DXing its meaning; however, here we need a light look at awards as a DX aid. After all, the screaming, hollering, frustration, and catastrophies that occur on behalf of the sport *should* result in something. And this something *should* help the DXer. Here then is how awards help us.

First, no award can help by being out of sight. Your DX awards should be on the wall! This way you can gaze upon them from time to time and smile. As with art, special, small highlighting lamps can be placed over or under them to further enhance their appearance. Elaborate frames can be used to give them an aura of importance. The awards themselves are designed with this in mind. Just look at the intricate borders of certificates or the solid feel of plaques. Lettering and terminology are also serious. Combine all of this with proper display, and DX awards can match even the fanciest diploma in providing a grand impression.

Once on the wall, these handsome certificates should be admired frequently. You should look up at them and say to yourself that it's all worth it. From time to time you should usher your wife or husband and children in to see them. (Do not include your spouse if there is the threat of divorce or physical violence.) You may want to bring in your dog or cat for viewing as well. They will, at least, offer no objections to your accomplishments.

After you've drawn all the sustenance and inspiration that you can from your DX awards, it's time to expand. After all, you've seen them enough and you may be getting a little tired of drooling over them. It's now appropriate to show them to other hams interested in DX. This can be done in casual invites ("Hey, come on over and see the shack") or you can send formal invitations with RSVP instructions. Heck, make it black tie. Notify the press.

When people come through, I suggest subdued lighting if the awards are specially lit or bright lights if they are not. Background music is most helpful in setting the proper atmosphere. (German march music is a nice little touch here.) Possibly even drapes could cover that portion of the wall that holds the awards and these could be ceremoniously drawn at an appointed time. And everyone will oooh and aaah as they see those awards. You just wait.

What? They didn't gasp? They didn't shake your hand and congratulate you? Someone in the group had a 330 country total?! You have how many? 225? That's a lot, you said? Then someone rolled on the floor laughing? And someone else gagged, spilling champagne on your QSLs? My, my—how awful. And then your wife came in screaming? She threw Cool Whip all over your equipment? How messy. What? Your neighbors came over? They heard the commotion and started attacking you about RFI and TVI? They stormed your tower?! Cut the coax? Tsk, tsk. Unbelievable. Your dog got excited? Bit you, you say? Dreadful. What did you finally do? Ahhh—you put the awards away. That started it all, huh? I see. What's that? Respect for the mysteries of DXing? Yes, I think I understand. There is a lot to learn.

DX CLUBS

Test. Answer the following multiple choice questions to check your knowledge of DX clubs.

1. In general, DX clubs are:

   ☐ Warm, friendly organizations that greet newcomers with open arms.
   ☐ Cold, snobbish organizations that shun people with low country totals (under 290 for example).
   ☐ Pseudofriendly groups that pretend to be nice to someone but then blackball him (while laughing behind closed doors).

2. The Purpose of a DX club is to:

   ☐ Give DXers a forum for bragging.
   ☐ Foster DX fellowship.
   ☐ Assist others in DXing.
   ☐ Sabotage others in DXing.

3. Most DX clubs have monthly meetings. The purpose of these meetings is:

☐ To drink martinis.
☐ To have DX discussions.
☐ To drink beer.
☐ To plan and execute DX-related activities (examples: DXpeditions and contests).
☐ To eat pizza.

4. To get into a DX club, one must:

☐ Have some money.
☐ Have some connections in the club.
☐ Have some self confidence.
☐ Be a DXer.

5. In comparison to other Amateur Radio clubs in a given area, the DX club is:

☐ Above the others in collective IQ.
☐ Below the others in collective IQ.
☐ The most sane.
☐ The most insane.

6. When members of a DX club get together socially, they:

☐ Stand tall, smile broadly and talk assuredly.
☐ Laugh.
☐ Discuss rigs.
☐ Show each other their rare DX QSLs.

7. When a guest is attending a DX club dinner meeting, it is appropriate:

☐ For a member to spill a drink on the guest.
☐ For the guest to feel strange.
☐ For no one to talk to the guest's wife or girlfriend (unless she is very pretty).

8. When a guest is attending a DX club dinner meeting, it is inappropriate for the guest:

☐ To ask a silly DX question, bringing down howls of laughter.
☐ To talk in a deep, impressive voice like real DXers.
☐ To get sick at the banquet table.

9. When trying to get into a DX club, the most certain way to make a *bad* impression with the members is:

☐ To tell tall DX tales.
☐ To claim to be running high power—1200 watts.
☐ To verify you have China worked to a club member who hasn't.
☐ To show a member your rarest DX card: ZP or ZS3, for example.

10. If you are accepted into a DX club, you should:

☐ Curse and swear.
☐ Jump for joy.
☐ Go to confession.
☐ Hide from your spouse.
☐ Deny it to your friends.

How did you do? Oh—you need the answers. Well, with tongue only partially in cheek, you could have checked off just about every answer because each one, more or less, has some accuracy or relevancy. Of course, we know it's not *really* right for a DXer to spill a drink on a club guest. On the other hand, DXers do have certain priorities or special privileges that allow them . . . oh well, better not stir that up.

The truth is that DX clubs *are* superior and, despite this, they are sometimes resented by the ham community at large. I simply cannot understand this attitude. There is, however, little that can be done as some hams just don't accept the importance and nobility of DXing. For those people there is little hope but how about the aspiring DXer? What does he or she need to know about DX clubs? Should one join? Are there any advantages to membership? Let us explore these questions.

DX clubs are, in certain instances, reminders of the old school or private club atmosphere. This attitude is, at first glance, obvious but it is also much less important than is generally understood by the outsider. In addition, in certain locales the DX club is downright local and open to all. However, all true DXers consider themselves special, and the tone or mood of even the average DX club usually reflects this. Don't for one minute think that DX clubs composed of predominantly serious DXers think they're not better than the other local organizations.Even so, DXers like other DXers to be in their club and unless it is unusually restrictive, new membership is not only encouraged but is the lifeblood of the organization. For this reason, I encourage newcomers who are really interested in DXing to contact their local club and get to know some of the members. Here are some of the ins and outs.

Some clubs are open to membership and those interested must merely apply and acceptance is simply a formality. However, many DX clubs are by invitation only, and a new application must be voted on by the membership. Depending on the constitution, a certain number of members must approve the applicant, usually defined in the number of negative votes (e.g., no more than three dissenting votes). There can be another restriction. Some club constitutions put a cap on the active membership total, and if the quota is currently filled then new members may not be taken in no matter how popular they are. It is important to realize this latter point if you are considering looking into a DX club.

If the club operates on a vote-in basis, then the best way to get started is simply to introduce yourself to some of the members. This is easily done at a hamfest, other club meetings, on the air, or at some other opportunity. If you are a new DXer, go slowly. Do not rush things. Get to know some of the club members and let them get to know you. An opportune moment will arise allowing you to discuss membership opportunities. You may even be sponsored by someone outright, and, of course, if you are an old friend of someone in the club, you can find out about the essential requirements for joining. Most of this is obvious, but the newcomer should understand the difference between being invited to join a DX club and simply signing up for any of the broader-based local clubs.

Should you join a DX club? If you're not really interested in DX, don't bother. There might be some exceptions. Some clubs are large and have other interests (such as contesting). To the extent to which the club embraces other ham radio aspects your needs could be fulfilled; however, DX clubs tend to be more interested in DXing and the thrust of the clubs activities—meetings, repeater talk, on-the-air operating habits, etc.—are generally tilted accordingly. If DXing turns you *off*, then joining the local DX club because they throw a good barbecue each month is a bad idea.

If you are interested in DXing, then I encourage membership. The clubs offer practical assistance and are fun as well. DXers love to get together and swap tales, and other hams, who hate DXing, loathe this sort of talk so the club provides a useful outlet. In addition, real assistance in antenna installation and rig help is often available. And DXing help is also offered quite frequently.

For those who are ready, I offer the following tips. Usually a prospective member of a DX club must meet at least some of the membership somewhere along the line. This may be an informal gathering after which an offer to join may come or, after submitting notice of wanting to join, may require formal attendance at an official meeting. Whichever end of the spectrum is used in pressing the flesh, there are certain things that the prospective member should be aware of.

What to Do

Always answer yes if a member asks, "Wanna buy me a drink?" Buy the person two. This is most impressive.

Be open, relaxed, and candid. If asked about any secret love affairs, for example, be sure to tell all. If you're not sure how something like this relates to DXing, don't let on!

When you describe your station, always put it down. Say that you know it's not a *real* DXing station but you're working on it. Then smile broadly.

Dress according to accepted codes of your area. Examples: East Coast—L.L. Bean; West Coast—pseudo-Hollywood; Midwest—brown shoes; South—brown shoes with brown stains on them.

Learn the proper responses in a discussion with members. Some good stock phrases: "Gee, you really worked that one?" "Golly, on such low power?" "Man, what technique!" "Gosh, you've been around some *real* DX!" It is all right to salivate while expressing these sentiments.

If it's a dinner meeting and the member sitting next to you has a smaller steak than yours, insist that you switch. If you spot someone who you know is likely to blackball you, then tell another member that the guy is really a DX role model for you. It is OK to look gleeful and wave your hands in enthusiastic emphasis, but try not to accidentally knock over something or slap another member in the face.

Be generous and thoughtful. If a member wants to test your 2-meter handheld because he hates that manufacturer, let him do so even if it involves dumping beer into the touchtone pad. Such actions prove you're a nice guy when it comes to testing and that you really believe in your tone pad.

When a member recommends that you purchase the type of station that he has, always say you will do it. When another member tells you to buy *his* type of rig also agree to do it. Whenever others recommended more choices, always agree that it's your next purchase. Do not be worried about discovery. The only way this could happen is if the particular members discuss among themselves what they've told you, but at this stage there will be violent, physical arguments which will eclipse your relationship with them. Just walk away from all this saying, "Yes, of course."

When a member invites you over to his house next Saturday always accept.Subsequently, do not complain at the top of his tower as you help install the new stack of monobanders.

What Not To Do

Don't go into the meeting with your country total flashing on battery-powered LEDs.

Don't even wear your ARRL DXCC pin with official total if it's less than 275.

Don't breathe excessively through your mouth.

Don't tell the members you have a Chevrolet.

Never tell the club that your spouse *likes* DX.

Although it can be anyone's objective to upgrade the station, never sound threatening or give clues as to what you might actually be doing. In fact, if questioned closely, nod uncertainly.

If there is a pileup, and in this pileup there is a club member, and if you need it and are tempted to turn your linear to compete with said member, DO NOT DO IT! (And do not ask why.)

Never accidentally break the private and exclusive DX repeater and if you do, die.

Do not challenge a member about his current, secret DX knowledge saying that, you too, read it in the bulletin.

Never give an established DXer advice as a newcomer. Even if correct, even if helpful, even if possibly decisive in a critical DX situation, it will not be believed. And if proved right afterwards there will be anger. Then, you might not get in the club.

Do not buy an Alpha before you get into the club. (Or, if you do, say you're using a Heath SB-220.)

Now that you're familiar with the nuances of fine DX behavior you might need some advice on how to fill out a club application. Here is a typical application form with tips and explanations for properly providing the information sought.

Name: (Be honest here; provide your name)

Call: (OK to give this too)

Country total—worked: confirmed: (This is vital information. Be accurate.)

Equipment: (List the minimum; don't sound excessive, like some of those California people.)

Antennas: (Again, be basic. If you have an antenna farm don't list everything.)

Bank: (Be factual here)

Signature for banking: (Reproduce this accurately here)

Worth: (To the penny)

Transportation: (Mercedes, BMW, Jaguar, etc.)

How much do you like DXing: (A good response: I like DXing so much I'd go to Kingman Reef myself to put it on—and pay for it!)

What sacrifices would you make on behalf of DXing? (The following is a nice answer: At the very least I would sell my personal possessions, including family, to continue DXing.)

Do you play golf? (Firm answer—no!)

If your neighbor's house is on fire and it is your turn on a list to make a DX contact, would you call the fire department or go ahead and make your QSO first? (Don't answer this. Write in a comical tone, sort of laughingly, that you know exactly which action has priority.)

Why do you want to join this DX club? (Good response: The members of this club are *real* DXers and I want to be associated with *real* DXers—*really*.)

Write a short paragraph on what DXing means to you. (Typical excellent responses: DX is like a warm fireside in a steelmill and like the delicacy of fine walrus skin. It is like the sun, the moon, the stars and great fishing ponds. It is the rush of excitement, the steeling of nerves and the forging of willing and even uncooperative elements of the inner and outer spiritual zones. And there is more!)

Are the following DX countries spelled correctly?

Fernando de Norona
Gurnsey
Tadzihk
Lacadive Island
Afganistan
Luxemburg
Madera Island
Leichteinstein
(You'll have to check these out yourself)

Match the following prefixes with their DX country.

T4	a.) Japan
XQ	b.) Somalia
HW	c.) Venezuela
CZ	d.) Yugoslavia
T5	e.) Cuba
4M	f.) Canada
4N	g.) Chile
8J	h.) France

(Tough, eh?)

That does it. If those tips don't help you get in a DX club nothing will. Use this valuable information wisely.

Now that we've poked a little fun at DX clubs, it's hardly necessary to tell people how to behave in order to be accepted into one. All this kidding around might confuse the issue as to just where I stand on DX clubs per se. I have mentioned earlier in this book that I am a member of one, and for emphasis let me state that I believe in the idea enough to have served as president and vice president of my club, the Delta DX Association. So, fun aside, my commitment to the DX club concept should be obvious. I heartily endorse new DXers exploring opportunities for membership.

7

Loose Ends

Next, we look into often overlooked ways to increase your DX capability. For example, did you ever have any foreign language training in school? If so you would be surprised at what you might remember—or potentially remember by brushing up—and how effective this is in DXing.

QSLing strategy will also be discussed below. This is no small part of DXing and must be vigorously carried out in the early stages (when it's tempting to quit because of the large numbers involved) and delicately executed in many rare QSL situations. Some things are as simple as enclosing a brochure about your city or state. Imagine how interesting this is to a foreign ham. A little touch like this can highlight your QSL and encourage a response.

Unusual QSL opportunities will also be explored. For example, many DXers came to the sport later in their ham lives. In such cases there may be some QSOs in their old log books with countries that are difficult or impossible to get now. For example, Libya (5A) and Ethiopia (ET) used to be quite active. That's right! Hams active as little as 10 to 12 years ago could have QSOs with these stations tucked away in their logbooks. There are surprising ways to get cards for these old QSOs, and the percentage of return can positively amaze you.

Of course, on the more serious end, such topics as the art of *receiving* QSLs will be discussed. When you've gotten a good one, bragging rights are in order, but it must be done with style and subtlety. For example, you should not crash into the DX meeting blubbering, "Look, Look! I got my YI card!" Rather, with glass of wine in hand, one should calmly state, "Oh, by the way, the YI card came in." Then turn blandly turn away and pick up a caviar hors d'oeuvre.

DXers are sometimes accused of valuing a QSL more than the QSO. These so-called purists feel that DXers only go in for the cards while it is the QSO that is more in the ham spirit. This is an old topic and it shall be given some consideration.

Many hams who are not contest-oriented frequently get disgusted when a major contest is coming on and avoid operating during them. DXers so inclined must learn to abandon this attitude, particularly in the early stages. Contests bring out good semi-rare and occasionally moderately rare DX and simply can't be ignored. Tips on contesting will be discussed.

You should appreciate the importance of this chapter. Understanding QSLing alone is a vital part of DXing, and if you are going to participate in an awards program, this phase of the hobby is essential. One other thing of interest to DXers is that the little points mentioned in this chapter are often overlooked by the *casual* DXer. Sometimes a significant DX accomplishment can be achieved by employing some of the ideas in this section. For example, by putting in a few hours in some old and new *Callbooks*, I tracked down a ham whom I had worked in C9 (Mozambique). I will relate the essentials of the story in the appropriate section in this chapter to assist others in their potential search for oldies but goodies, but the main point is the value of such sleuthing if one gets a card. If one gets just one such card—or QSO from this other advice—then this whole chapter is worthwhile. It is the extra step that many do not take that can make you a better DXer.

FOREIGN LANGUAGES

This is a "real sleeper" in DXing and a weakness in the amateur community. Few people realize how much the population of a given country, including the hams, are appreciative of anyone who makes a reasonable attempt to converse in their language. Enormous bridges can be successfully crossed here, yet the number of hams who fail to utilize this area is remarkable. True, no serious DX career is ultimately impeded by the lack of a foreign language. Most countries can ultimately be worked; however, equally true is the fact that a given foreign language capability opens doors that most never appreciate.

Before I get started in the specifics of learning another language (which is not all that challenging for the *basics* and which many hams already have in their memory banks) let us cover a few situations where another tongue was helpful. In this way, the value of a little effort can be appreciated and will hopefully be motivational. To do this, I have to rely on personal experience. The language involved is German. (Spanish and French are more commonly taught in our high schools and universities and probably more hams have a working knowledge of these Romance languages than German.)

How useful? Plenty. As stated before, those DXers who have not used this subtlety will not fully appreciate it's value. It's easy to scoff since English is the international DX language, and the "foreign language advantage" usually does not affect the critical difference between getting and not getting a country in the long run. And yet, to those who know, it can make life much easier.

My first YK contact (Syria) was with an Austrian station in the Golan Heights. As always, much goes on in the Middle East causing suppression of ham radio and there was no exception then; in fact, though Syria was very noisy on the political front it was all too quiet on the ham radio front. The pileups for the particular YK were formidable. In the thick of it, I called in German phonetics. After a few calls, the DX station stopped the pileup and singled out the ham who was calling in his language. Easy contact. Another example was 5N. A West German operator was there. In a heavy pileup, I was singled out specifically again. The same with XT. Bragging? Hardly. This isn't DX *skill* by definition. This is just another edge which *many* have. Did you study French or Spanish or another language in high school or college? If so you have at your fingertips the ability to do something more in QSOs. If not, you still have the ability to learn some basics, if you desire!

Besides being a DX edge, a good working knowledge of a language opens other special doors such as ragchewing, new friendships and, almost invariably, invitations to visit if ever in that country (and these are the *sincere* type of invitations in which you will be extended considerable hospitality). Consider ragchewing; I remember a fairly long QSO in German with TN8AJ. Many serious DXers would have given up their first-born child to have had that QSO!

For DXing in general, however, a strong ability in a language is *not* necessary. What is needed is the ability to say the alphabet, count a little and know a few general phrases. There are some aspects of this that should be paid attention to, however, to more effectively use the basics of a language. If these aspects are not studied, you can actually set yourself back in trying to make yourself understood with foreign DX.

The first of these aspects is pronunciation. This is the biggest mortal sin of all. Far too many Americans don't take the time to learn how to properly pronounce the foreign language they are speaking. I am not trying to send the whole ham community to language school but certainly one can learn how to say the alphabet properly in another language. Too often hams give their own call in a foreign language with which they are somewhat familiar but they slaughter the pronunciation. This *increases* the difficulty of the DX station (who usually understands good English pronunciation) in his ability to copy your call. One can then frequently hear the DX politely switch to English in order to clear up the QSO.

This might turn off some DXers and scare them away from trying a foreign language. I hope not. For one thing, some foreign hams switch to English no matter how well you are speaking and you can't let this discourage your efforts. There is nothing you can do about this; however, it is still important to check your pronunciation and make sure it is reasonably accurate. Don't add to the confusion by saying it wrong.

The next aspect is to select the phrases you want to use and to understand precisely what they mean. Unintentional misusing of phrasing and meaning only adds confusion. Along with this, make sure the phrases you want to use are in proper form with correct tense. What *sounds* goods in another language may still be botched up on the other end if improper grammar is employed. If you want to set up a future schedule with some DX station it doesn't do to say the equivalent of, "I had met you tomorrow at 0400."

The next aspect is not to get ahead of yourself. If you really don't know the language well, but are properly versed in pronunciation and a few phrases, then exchange only the basics. In the DX station's language, say each of your call signs with report and a few pleasantries, then switch back to English. Going further than your ability adds to confusion yet it is hardly an earthshaking matter. It is usually just a little embarrassing or awkward as you admit you can't go further. Ironically, good pronunciation sometimes leads the DX station into thinking you're more fluent, but this misunderstanding is usually flattering!

Now, how do you go about refurbishing or learning some language ability. For those of you who have studied a language at any time in your life, you are three-quarters there. Way back in your brain is the ability to converse basically. Though the mind may be very rusty, once you start you will be surprised at how quickly things return. If you have any old texts, take them out of the closet and begin to use them. If you have none, a beginner's book and pocket dictionary are available in many standard bookstores or through university or college bookstores.

There are also several cassette series for basic language study. These are excellent and really help to get pronunciation straight. Remember, for basic QSO exchanges and pileup calling, there isn't really much to learn. After getting the alphabet and numbers zero through nine, you have most of what you need to call and answer a station correctly.

Phonetics are easy. Just use standard ones. See Fig. 7-1. It is not necessary to use Bach or Beethoven for Bravo when calling in German.

One other point to remember. In a pileup, call *often* in the alphabetical language itself. Use phonetics only as a backup for clarity. It is the actual call letters in the language that stand out and attract the DX. Hearing his native tongue is always a drawing card for the DX station, particularly if he is visiting another country on a DXpedition or other assignment.

For those who have some advanced knowledge of a language I urge you to develop it or rehabilitate it if it has been dormant. Besides meeting foreign hams on a much more in-depth basis, it is also useful for DXing. I recall several times talking to different German list-takers in advance of a list operation. Since we spoke in German, my call was more firmly implanted in their mind and—

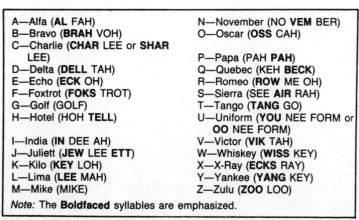

A—Alfa (**AL** FAH)
B—Bravo (**BRAH** VOH)
C—Charlie (**CHAR** LEE or **SHAR** LEE)
D—Delta (**DELL** TAH)
E—Echo (**ECK** OH)
F—Foxtrot (**FOKS** TROT)
G—Golf (**GOLF**)
H—Hotel (HOH **TELL**)

I—India (**IN** DEE AH)
J—Juliett (**JEW** LEE **ETT**)
K—Kilo (**KEY** LOH)
L—Lima (**LEE** MAH)
M—Mike (**MIKE**)

N—November (NO **VEM** BER)
O—Oscar (**OSS** CAH)

P—Papa (PAH **PAH**)
Q—Quebec (KEH **BECK**)
R—Romeo (**ROW** ME OH)
S—Sierra (SEE **AIR** RAH)
T—Tango (**TANG** GO)
U—Uniform (**YOU** NEE FORM or **OO** NEE FORM)
V—Victor (**VIK** TAH)
W—Whiskey (**WISS** KEY)
X—X-Ray (**ECKS** RAY)
Y—Yankee (**YANG** KEY)
Z—Zulu (**ZOO** LOO)

Note: The **Boldfaced** syllables are emphasized.

Fig. 7-1. ITU standard phonetics.

wonder of wonders—I made the list. I asked for no special favors nor would I have received any. My call stood out, however, psychologically as well as RF-wise.

An excellent language series specifically for hams is put out by Carl Sletten, W1YLV, in Acton, Massachusetts. His text and accompanying cassette tape are targeted for ham radio and are superb. The only thing is that there is very little grammatical explanation for the sentence structures used and even though it includes beginning material a background in the language is most helpful. Even so, the specific amateur and DXing nature of the program makes it an excellent aid for those wishing to increase their language proficiency in this area. The ARRL also has a Spanish-language ham radio course available. It is specifically ham oriented for those seeking to broaden in this area.

Use of a foreign language in amateur DXing is fun and useful. As previously pointed out, it is not essential in DXing, as sooner or later a given country can always be worked. Those who don't have this tool, however, don't fully realize the valuable asset that it is. Only a minimum of effort is required to develop some language facility and, if handled properly, is always appreciated on the DX end.

QSLing--A REAL ART

QSLing is indeed an art. At its most basic it is the simple exchange of cards between two hams. At its most sophisticated, it is complex negotiations with difficult DX operators and the careful guidance of your card and money or International Reply Coupons (IRCs) through the shadowy network of the overseas postal system. Then there's the Russian situation. All DXers work all the Russian DXCC countries, and many know the frustration of confirming all those countries. In fact, it is usually a red-letter day (so to speak) when someone gets his final Russian card. What to do? Read on!

Let's start with basic QSLing. I have seen advice in magazine columns that stress bureau use. While the bureau is valuable, I recommended QSLing direct in many instances. Each country is one unit in the total and each card is important. Don't take anything for granted. Of course, if one is at or near zero then it is unnecessary to panic and send a QSL direct for a VE card. The bureau will do nicely enough. Actually, most of the first hundred—the *common* Central and South Americans, Asians, Europeans, and Russians—

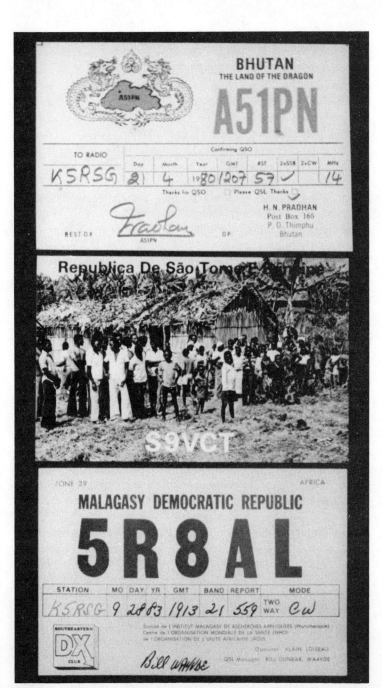

Semitough. This is not your everyday stuff but not impossible either.

can and should be achieved with bureau mailings to save time and money.

The best and easiest method for high numbers of cards is through the ARRL Outgoing Bureau System. One of the benefits of League membership is being able to use the ARRL Outgoing QSL Bureau to conveniently send your DX QSL cards overseas to foreign QSL bureaus. Your ticket for using this service is your *QST* address label and just $1 per pound. Your cards are sorted promptly by the Outgoing Bureau staff, and cards are on their way overseas usually within a week of arrival at ARRL Headquarters. More than four million cards are handled by the Bureau each year.

QSL cards are shipped to QSL bureaus throughout the world, which are typically maintained by the national Amateur Radio society of each country. In the case of DXpeditions or active DX stations that use U.S. QSL managers, a better approach is to QSL directly to the QSL manager. The various DX newsletters, the *W6GO QSL Manager Directory,* and other publications are good sources of up-to-date QSL manager information.

Instructions for sending your cards are available from ARRL Headquarters for a self-addressed, stamped envelope (SASE). I recommend this method, as they make it easy for your cards to get to the various bureaus of the myriad of countries you are working over and over. The other alternative is to send your cards to each country's bureau yourself. This is of course tedious and almost as expensive as direct QSLing except that one does not have to enclose IRCs, money, and envelopes for the return card. There is a time when going direct to a bureau is useful, however. Suppose, for example you have a few Russians out and not much else. Or, say you've knocked off a half dozen HA contacts but, as luck sometimes has it, you've gotten no QSLs after a long wait. In these cases, mailing your cards directly to the appropriate bureau is faster and easier than the extra step through the ARRL. Remember that addresses for the various bureaus are in the *Callbook*, listed together in the front section.

QSLing direct is easy but a few points are in order for beginners. Always remember you are trying to make it easy for the DX to send a card to you. Therefore, always enclose an SASE for stateside QSLing when dealing with U.S. managers. Overseas QSLing involves sending a self-addressed envelope, usually *not* stamped. IRCs are usually all that is needed for common countries. Remember, send enough IRCs for *airmail* postage (chart for this is in the *Callbook*). The point of direct QSLing is for fast, efficient return. If you're going to send one IRC for surface mail, then you might as well go through the bureau.

Whom should you QSL direct to? Once you get to the level of countries like Guernsey, Isle of Man, Faroe Island, Egypt, Ivory Coast, Niue, Norfolk Island and Gibraltar—to name a few—I think direct is useful. These are not "everyday countries" and, in my opinion, careful QSLing will result in a higher percentage being confirmed sooner. This saves you the hassle of getting back in pileups which one tends to do for a previously worked but still unconfirmed country. And, in the end, you will find that some of these countries will not bother with the bureau or it takes ages for them to do so which, in either case, results in one finally QSLing direct after all.

For the next level of countries—somewhat rare and rare— QSLing becomes more specialized. There are two types of QSLing procedures in this category: those to DXpeditions and those to hams living in the DX country.

In the first instance, QSLing for DXpedition contacts, sometimes a "green stamp" (dollar bill) is requested in lieu of IRCs. Money is, however, almost never an absolute requirement for getting a card. At any rate, depending on the importance of the country and the probable cost of putting it on, I usually enclose a given amount of cash, IRCs for return airmail postage and a self addressed envelope. If it's to a stateside QSL manager, the enclosures are the money as designated and an SASE. This covers by far the vast majority of DXpeditions. Usually sending American money (even to non-U.S. QSL managers) is no problem as the QSLing addresses are most often in countries where this does not present any difficulties.

In the second type of QSLing for semi-rare and rare countries— those in which the hams live—there are some tricks you should know. The first thing you have to deal with is the operator. Some QSL aplenty. Others do not or are very slow. Then you have to deal with the country itself and its postal and regulatory policies. Since the latter is more complex, let us deal with it first. You should always enclose the self-addressed envelope and some means to assist payment for return airmail postage. Green stamps are desirable and often appreciated by DX operators; *however, this is not always the case.* Some countries prohibit transfer of money— particularly foreign money—by mail. It is a crime, and the recipient can be held responsible. You are not doing the DX station any favors in such cases. Even if the recipient does not personally get into trouble, the letter can be opened, the illegal cash confiscated and the remaining contents seized and destroyed. The countries which, in any given moment, are active in doing this are usually discussed in the bulletins. It is important to scout various information sources at QSL time for those countries likely to be involved. One should then turn to alternative payment for return postage; i.e., IRCs.

That's not all. IRCs are also worth money to postal or other personnel and thus, even if it is not official policy to open letters to hams, outright theft "justifies" so doing. The end result is the same. The letter and your QSL are destroyed to eliminate any evidence. What to do? The first line of defense is the envelope. Hams love to be *hammy* and put Amateur Radio Station such-and-such and Chief Operator so-and-so. Avoid this like the plague. These are the red flags which signal the various officials enroute to selectively open that envelope. The DX operator's name alone may be targeted if he's one of a few amateurs or perhaps the only one in a particular country. Don't add to the identification by putting call signs.

Make the envelope bland. Address the person by name only. Avoid using Mr. or Mrs.; those earmark it as foreign mail. A typewritten, addressed letter is an alternative, because it appears more business-like. It is also clearer than script and thus appreciated by postal officials, and the majority of hams don't type their envelopes. Some mail has to pass through, so you increase the chances that yours will be one of them. The final concept—from the mailing standpoint—is the enclosure of the DX's (or manager's) country's stamps. This is, selectively, a good idea. Stamp stores carry almost every country's current stamps. Though it is easier to misjudge what is required for the return postage, a dealer can usually help calculate it. At any rate, it's hardly a killer in price and is a good way to encourage a timely response. Put the stamps on the self-addressed return envelope. This makes it easier to mail and reduces the occasional temptation to use the stamps for another purpose.

Next, you have to deal with the DX station who may or may not be a generous QSLer. I believe it is a nice little extra touch to enclose something which may be of interest to a foreign ham. I usually include a small brochure about my city, New Orleans, and have found the response to be remarkable. I have had special city or state stamps sent to me or friendly letters or other memorabilia and it's all been wonderful. The DX hams love it and so do I. The fact that the QSL card is also enclosed is, of course, moderately incidental!

Finally, there is the case for special enclosures. Some DX hams collect certain things: stamps, gold coins, antique Rolls, and so on. Won't you help by sending along a few Krugerrands? Seriously, though, many DX hams are collectors of various tidbits, and helping them can help you. Probably one of the more common hobbies DX hams are involved in is stamp collecting. An enclosure of some nice stamps for the DX station's collection is a plus. The tips about which DX operator collects this or that usually circulate on the air and are mentioned sometimes in the bulletins.

The preceding paragraphs cover the major points of DX QSLing. There are two other special cases to be explored: the Russians and unusually tough, restrictive countries or operators.

Virtually all QSLing with Russians is handled through the bureau. All cards are sent to the famous Box 88 in Moscow. What about direct QSLing possibilities? Much of the preceding information applies here, but there are some special points. First, some of the rare Russian prefixes (Franz Josef Land, for example) are put on by special operations, and there is the equivalent of a manager in Russia. A direct QSL to the manager results in a more complete and rapid return rate. Frequently, this information is not announced on the air by the Russian DX operator. It is published in the bulletins, however, and is usually known in DX circles and discussed on the air (see Chapter 5). One may enclose Russian stamps (available at stamp stores), and I really recommend this. Put them on the self-addressed return envelope. I usually throw in a few extra ones for the ham's own use.

There is another aspect to Russian QSLing, that of the regular ham or club contact as opposed to a special operation with a manager. At some point, after having worked all the Russians, it will dawn on you that a certain percentage of the cards are going to be very hard to get. Direct QSLing can result in getting some of those cards. For this you have to use the *Callbook*. Unfortunately, the addresses are incomplete. Usually the name of the operator and the town and district are given but not a specific address. The same is true for the radio clubs, of which they are many. The name of the club is given but again not the address.

Fortunately, this is sometimes enough. If the individual ham lives in a small town, it may reach him. In the case of a club station it is easier for postal officials to get mail to these official government institutions. I recommend using Russian stamps again and enclosing that little something extra (brochure on your QTH for example) with the self-addressed envelope. The results can be surprisingly good. When I was down to my last confirmations needed after *years* of waiting, I sent three letters as described. Three QSLs resulted, and this was after a considerable number of attempts through the bureau.

There is one final thing about Russian QSL cards. There are some hams (in and out of Russia) who have very good personal con-

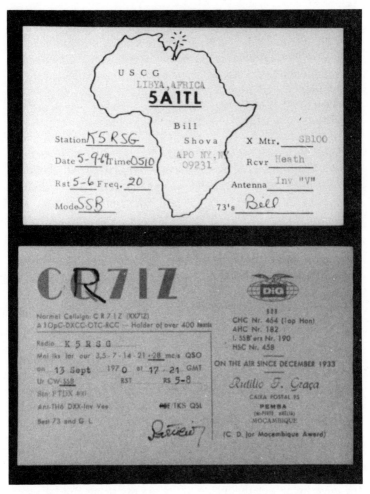

Easy yesterday, hard today. Libya and Mozambique were quite common in years past. This is not the case now, though the political situation responsible for this scarcity could change.

tacts with the "right" Russian hams. Sometimes they can get you your card when all else fails. For obvious reasons, I cannot publish the information here as these people could be inundated with unreasonable requests. It takes years to get all the Russian cards, and impatient hams cannot and should not use this route prematurely. If you are a real DXer and are careful and patient in your methods, you will get most of your Russians. If you finally do need assistance you will, at the right time, know who to contact.

Some might ask is all this worth it? Why not get on and work the particular Russian countries over and over until you get the card through the bureau. It's not that easy. True, some hams have the right combination of luck and fly through the Russian confirmation problem. However, this is the exception, not the rule. And the solution is not to simply keep working them although every DXers goal should be multiple contacts.

First of all, some of these countries are rare so you can't work a bunch of them. Often, only a handful of stations are on in some of these places. When you work the country again and again, you are contacting the same stations which did not QSL in the first place. Second, the problem is, ironically, often inverse to the number of contacts made. Not infrequently you get the card from one contact with one country and don't get a card from multiple contacts with another. This same can be true within even one country. After one contact with Armenia I got my card while a friend of mine took over a year longer to get his, even though he had about a half dozen contacts.

As can be seen most DXers will encounter some degree of the "Russian problem." Therefore, these QSLing techniques are worth it and are uniquely rewarding. It is very satisfying to use a somewhat off-the-beaten-track method to squeeze out a rare card.

The second special case of difficult QSLing is that of unusually tough and restrictive countries. Though fortunately rare, this can provide one of the most unusual QSL challenges to be found. Typically in these countries, on-the-air activity is tenuous at best, but the QSL problem is even worse. Although DXing insiders may profess to "know" what the cause of the problem is, probably no one has a complete handle on this complex picture. There are several potential causes: the hams in a given country are dilatory about QSLing or the incoming letters are being officially seized by the government; or the incoming letters are being stolen. These same potential causes can apply to any country of course; however, whatever the cause, the total application of QSL restriction is a rather severe action which has a blanket effect on the world amateur community.

The causes of this problem—and the solutions—are interesting. Though I don't know the personal energy of the operators involved in problem countries, it seems to me that they would QSL if they could. The intermittent surges of QSL activity that arise in these countries would seem to support this. Of course, they could let the QSLs accumulate until they feel they *have* to send out some and they've collected enough "enclosures." This theory would be believable on the basis that they must QSL occasionally or they would lose their credibility. Then QSOs and QSLing would stop altogether. Another scenario is theft; however, the QSL spurts include previous contacts, so if theft were the sole motive, incoming QSLs would disappear and not be answerable. This therefore seems unlikely.

The only real causes left for massive QSL suppression are therefore probably governmental. This is hardly an astute observation. Yet conditions leading up to suppression (and the easing up of same) are interesting and what is said here is applicable to any country though the reasons for restrictions may change. Thus, the solution to the problem is the same as the cause—the government and the complex politics behind it. This is, of course, a most unfortunate course for amateur-related activity to be impacted upon, but the essence of political situations in many countries dictates realities far worse than amateur considerations. In addition to governmental meddling, there are other foul-ups of various types that no one can control or prevent. In short, it's an imperfect world out there.

From the standpoint of QSLing in these cases, there is obviously no control which the DXer can exert. However, keep your envelopes as bland as possible and look for specific address instructions for these special situations in DX bulletins and on the air as well. What with government sentiments being highly variable the last thing you want is to have some postal official—uncertain of which way the political winds are blowing that day—select your letter to be "purified."

One lasf ditch trick is to send your letter through another country if you have a friend living there. This avoids the U.S. "look" on the envelope which sometimes tempts postal officials for economic or political reasons.

QSLing can be a simple matter, and for the average ham it usually is; however, for the DXer, cards are very valuable and essential for participating in the various award programs. Thus, the measures described have to be employed from time to time. I hope that this review of QSLing techniques has been thorough and has "covered all the angles." This has all been the result of years of learning both on my part and the part of others. Careful consideration of the points raised will certainly cut down on the hassle for some who are trapped in the QSL maze.

OLD QSO, NEW QSL

For those of you who have been around for a few years but came to DXing late, there is a potential for some good QSLs. Drag out your old logbooks and check your contacts. A lot of countries were on that are off now or if on can still be tough to get. If you worked 'em then you had a QSO and it's as good as one now. The problem is that during those years you may not have taken DXing seriously and you didn't energetically QSL. Tch, tch. You're going to feel real bad when you had that easy XW (Laos) contact and now find that the current Lao People's Democratic Republic isn't very democratic when it comes to ham radio activity. Well, all is not lost. Many cards can still be recovered and it is very much worth looking into.

The first thing to do is to go through the logbook and find the QSOs. Next, get callbooks from the period and new ones as well. Check for QSL managers in the log. I was surprised at how many managers—after eight to ten years—were still listed in the current callbooks. All of the countries I found in my old logs were not rare but many were tough to get. They included the following: 9X (Rwanda), 7P (Lesotho), TA (Turkey), 5V (Togo), TJ (Cameroon), FL8 (now J2, Djibouti), JD (Ogasawara), EP (Iran), 9V (Singapore), ET (Ethiopia) and CR7 (now C9, Mozambique). Though some are not rare all are good ones, and ET and C9 are very rare. These contacts sat in my logs unknown to me for over a half year after I turned serious about DXing. Well, if *I* had these, think of what you might find if you were active years ago.

The results: Of these eleven QSOs, I thought all were worth sending a card if I could get QSL information. In the end, I got seven confirmations including Mozambique which alone made the whole effort worthwhile. QSLing in many instances is easy. If managers are currently listed in callbooks, send them a letter and card. You will be surprised how many retain logs and cards just for such occasions. (None of the DX stations themselves were currently listed, so direct QSLing was not possible. Nevertheless, be sure and check this out for any of your old contacts.)

Despite managers who are still around in some cases, in other instances it's not so easy and you have to be more diligent. I will outline one story. I had three contacts with CR7 (C9) in 1970. Using a 1969 *Callbook,* I got addresses for the three different stations I had contacted and, not expecting much, I sent letters with cards to Mozambique in the hopes that maybe one of the hams was still there with at least his logs and some QSLs. The letters were eventually returned undelivered. Three strikes, but not out!

Without even waiting for these returned letters I also decided to try another tack. It seemed logical that many would flee to Portugal in the middle of and after the revolution in Mozambique during which time ham radio activity came to a halt. Using the hams' names from the 1969 *Callbook* I then searched, name by name, the Portuguese listings (over 2000 calls). (Note: The *Callbook* publisher can sometimes assist here with its computer listing.) Of the three hams formerly active in C9, I finally found one who had become licensed in Portugal. I wrote him and struck gold: He verified my old contact and sent a very nice letter, in part unhappily describing how ham radio came to an end as a result of the political situation there.

This story clearly illustrates the value of a little sleuthing with old contacts. First, it was intriguing work, and second, it was fascinating to renew contact with a ham I had talked to many years ago and to learn what happened to someone caught up in a major political maelstrom.

Before closing the section on QSLing, it is best to cover an important trait: patience. Exchanging cards is slow business. As your totals climb, the type of stations worked causes slower QSL rates of return. This is because of several factors. Exotic countries having few operators results in a small number of hams serving the entire DX world. Their QSO and QSL chores are thus demanding and time consuming. In addition, postal service in these countries is often slow and inefficient.

In the case of DXpeditions there are also long delays. Returning and disbanding from the DXpeditions takes time. QSL design, layout, and final printing are usually not speedy. Then, as people are returning to their regular lives, massive QSL duties from an effective expedition takes considerable time to get geared up for. Although some delays result from unnecessary factors, most are legitimate and involve problems of logistics inherent in DXing.

It is important to know how long to wait so that one does not prematurely initiate a second inquiry. Experienced DX ops usually are less than pleased with these premature requests because they add to the already formidable mail problem (receiving, sorting, coordination with logs, etc.). Some will actually delay your card if the request seems unusually out of line or they get irritated.

The worst situation, however, can happen with the truly uninformed DXer. On too many occasions, nasty letters are sent to the DX station and this can jeopardize not only that card but future operations as well. One such exchange surfaced in a DX bulletin. The irate DXer's letter and the response from the South American DX op (who had gone on a DXpedition) both appeared in print. The poor DXer. It seems he sent *four* requests in under 120 days after the DXpedition. The last one, as reported in the bulletin, was rather caustic. The South American ham outlined the problems involved in QSLing in considerable detail. His rebuff was irrefutable. At the closing of the DX op's response, he clearly stated his feeling toward ever working that particular DXer again. Tch, tch. A black mark—and so early in DX life.

TECHNIQUES FOR PROPER, DISTINCTIVE RECEPTION OF QSLs

When your San Felix card rolls in, it's party time! The satisfaction of having proof that you did well in competition is well earned. It is therefore appropriate to notify friends and enemies of your triumph. However, *how* you do this is important. You can do it like Archie Bunker. Or like Laurence Olivier. Like everything in life, proper QSL reception requires thought and care.

Doing it right requires planning and patience. For example, never tell a fellow DXer that your BV card just came in when he's busy on something else. He won't be concentrating on what you are saying so the effect will not be the same. Instead, be patient, wait for him to say how he's looking for BV2B and then say, "Oh yeah, my BV card came in a few days ago." The difference as a result of careful timing should be obvious.

The ultimate goal is, of course, to do it in penetrating style. Real DXers love pain and they'll like you more if you sort of jab them with the news that you've gotten in a rare one. Always remember that it is expected and that, in turn, they will do it to you. And don't just limit this to those friends who don't have the card. It hurts even those who already have it in the fist, since this means you are moving closer to them.

Here are some tips. If you are in a DX club that meets on a

repeater, you can announce your acquisition there. Be sure to select a meeting or net when many members have checked in. Remember that this is the mass pain approach so you want as many people to be around as possible. Clubs that have a list of their members and what countries they have confirmed take updates on the net to add to the list. If this is your situation, you can save a few minor updates for those occasions when you have something really big. Then, as you announce the less critical cards, you lull your comrades into thinking that you don't have anything threatening. Then you hit 'em with the biggie just when it's least expected. For example you might handle it this way. You say, "I have the following QSL updates; SV Crete, KG4 Guntanamo, KH6 Hawaii and uhh, there was one more, what was it? Oh yeah, A5 Bhutan." Then immediately let up on the mike to turn it back to the net control. You can almost hear the moans at this instant.

When informing someone in person that you've gotten a good new one in, there are also special things to do. Many opportunities for verbal presentation come at DX club dinner meetings, conventions or just any ham gathering. Frequently, there is a social hour preceding the dinner or meeting during which times refreshments are served. As you are talking to someone let him take a sip of his drink and just as he is about to swallow say, "Oh—my YI card came in." This almost always will cause your companion to gag and he might even uncontrollably blow into his drink, splashing it into his face. It's good to carry a handkerchief around for this. After all, you want to be courteous. Other good times to pop the information out are when someone's eating soup or oysters. Be considerate. Known cardiac patients, for example, should first be asked to sit down and take a deep breath before breaking the news. This cuts down on medical accidents.

Another aspect of informing compatriots about good QSL receipts is your demeanor. The code word here is nonchalance. Always be low key. Here is an example. Let's say it's the ol' social hour again and you and a few DXers are standing around, talking earnestly about DX matters while waiting for a round of drinks. As the lovely waitress approaches with beverages for your group, she suddenly trips and the tray lunges toward you and the others. Without getting excited you should—in smooth, fluid, continuous motion—snatch your drink in midair, assist the young lass to her feet, reorder the refreshments, and then turn to the group to mention that after a little research you received a card from an old contact, 5A. You can add that what with Libya being what it is these days, this is rather swell. Then switch conversations. Say something like you're considering buying a Rolls. It is not necessary to say that you actually mean *sweet rolls*. Finally, calmly walk away from the group while all are standing there with their mouths open. Proper demeanor can go a long way toward adding to the impact of announcing a rare QSL reception.

Another aspect of QSL receiving is humility. Always downplay things a little ("Shucks, it was just a little ol' UH8 card. And a little ol' UM8. T'weren't much.") And show respect for others. ("I'm sorry because you didn't get through to Uganda, and I'm double sorry because the card for my contact came through today.") And try to be reassuring. ("Isn't this South Sandwich card pretty? When you get yours, I'm sure it'll be just as nice.") There is simply no end to the little touches you can add which round off your presentation.

Proper reception of good QSLs is a subspecialty of DXing. The techniques come quickly to DXers and hopefully this section will get beginners off on the right foot. Be sure and not make the mistake of using these recommendations too soon in your DX career. It simply won't do to spectacularly announce reception of a 9Y or CT3 card. People will chortle and guffaw. Wait till you get to a reasonable level and the cards have some significance.

QSO VERSUS QSL

A line is drawn across ham radio. On one side stand DXers who love QSL cards. On the other side stand hams who claim the QSO is number one. To some extent the debate is exaggerated but it crops up enough to be dealt with in a book on DXing. Those who feel that DXers concentrate too much on the cards and that the often brief DX contact is too short, argue that this is why serious DXing is not in the spirit of Amateur Radio. The QSO should come first in priority they say. Couple this thinking with the line, "I like to work DX but I don't bother to get the cards," and I'm sure you will recognize the total attitude.

Most hams like cards initially. Some hams become lackadaisical later. Criticizing a DXer's interest in card swapping may thus represent an excuse for laziness. Some have genuine dislike for placing a high premium on QSLing and are genuinely dissuaded from serious DX competition as a result. This is OK. To each his own. However, DX competition in various award programs *requires* QSLing. Given this situation, QSLing is an inherent part of the game. It is also the end result of a QSO and that part of the game which is visible and concrete. If that is to be criticized, so be it.

When it comes down to what is the most thrilling, the QSO or the QSL, I don't think many DXers will quarrel with the view that nothing beats the actual working of a rare DX station. That sense of triumph is when we find a given DXer at his most effusive. The card is exciting later, but nothing matches the moment of getting through on the air.

The non-DXer, of course, rolls his eyes to heaven at all this rompin' and stompin' on and off the air. And the DXer cringes in righteousness at what he sees beyond the QSO. Here's a short poem to summarize.

QS-L vs. O

To the optimist and the pessimist
  Life is very droll.
For the optimist sees the L
  While the pessimist sees the O
The pessimist thinks it's over
  When the QSO is through.
The optimist thinks it's great
  When he adds the card too.

The reason is not the first part
  The QS is the same.
It's the ending O or L
  Which adds the anger or fame.
To one it's so serious
  To the other so mysterious.
To each it could be nothing
  Yet each make it something.

Keats won't roll over on that, but perhaps a tiny point is made— different cups of tea and all that. How do we resolve these differences? It's simple. In the grand order of all things the primary point to remember is: DXers are right.

CONTESTS

Many DXers are also "contesters." Many are not. In fact, some hams love contests. Others love to hate them. If you are a DXer, you don't have to participate in a contest from the standpoint of winning it but as a source for new DX that simply cannot be passed up. The idea of turning off the rig during the weekend of a big contest to avoid the noise and confusion is misguided. If you are a DXer, you need to be in there, particularly at the beginning and middle levels of the DX climb.

There is a specific strategy for serious contesting. It is important to understand that this is *not* the same strategy for serious DXing in a contest. When looking for new countries, you do not need to worry about making a lot of contest points. The main goal is to cover as many frequencies as possible to check out the numerous pileups. There is much good DX readily available in contests, particularly the major ones. Some pretty rare stuff might be on the air. After all, many spots are activated solely for contests. Conversely, really rare countries usually do not participate in contests.

The nice thing about there being good DX in a contest is that the DX is also trying to work as many as possible. Thus, even if the pileups are big, the press to simply make contacts for high scores usually results in increased efficiency and high QSO rates. A contact is almost guaranteed with any effort.

Keep the search going. Put in as much (or as little) time as you want—time spent will usually be quite productive. In the big contests, pileups can be numerous. It can be easy to pass over something you need, so always search carefully. Something you need can be adjacent to something more rare, with the smaller pileup being somewhat camouflaged next to the larger one. Don't overlook seemingly unimportant pileups.

Contest are very fluid, with a lot of QSYing. When the DX station thinks he's milked a given frequency dry, he will move on. Therefore, never assume a group of frequencies you've already checked have nothing you want. Stations move around fast and what was barren moments ago may yield something juicy all of a sudden. Check and recheck are the watchwords.

You might think that there isn't much QSL productivity from contests. Wrong! QSLing by the DX stations is usually vigorous. They appreciate the contacts and are glad to QSL those who gave them points and need the cards. Occasionally, some of the the DX contest operators use the bureau even if you went to them direct. This causes delays of course and if you've sent IRCs, return envelope and possibly even a little money, it may be a bit frustrating. Remember, though, if the DX station did well in a given contest, there are a lot of cards involved. Most will respond, however, if you do so. Some QSL the remaining stations who send no card through the bureau, thus ensuring that almost everyone gets a card, requested or not. All in all, if you work some new ones in a contest you are very likely to get them confirmed.

As stated at the outset, contesting from the standpoint of competitive participation is not addressed because these subjects are outside of the scope of this book (see the *ARRL Operating Manual*). A serious attempt at putting on a DX spot for a winning contest score requires antenna, equipment, and operating commitments that are substantial. Stateside contesting, alone or with others, also requires considerable effort if high scores are the goal. If you merely want to concentrate on new countries, the approach described is both easy and effective. And, if you have avoided contests in the past, you can experience the fun of being a participant and the ease with which you can work many new countries.

8

Awards and DXCC

There are three basic purposes for working DX: 1.) the actual thrill of radio contact with distant lands; 2.) the competitive aspect of contacting a station in a competitive environment; and 3.) recognition. It is the third purpose that puts medals on the uniform, so to speak, and probably is the stimulating force behind much of what is controversial about DXing. Certainly one could not explain the desire to talk to a place like Kingman Reef on the basis of locale or distance alone. Rather, it is its value as a "country" and its correlated worth in award programs that stirs up all the fuss when it comes on the air.

Certainly DXing would exist if there were no awards or recognition. The pull toward talking distantly is too strong to be resisted in radio; however, with recognition comes goals and the big competitive push derives from this. This is inherent in all competitive activities, of course, and it is what makes playing the game worthwhile. Recognition is also the uniting factor of the sport, bringing DXers together in a variety of forums.

Obviously, recognition cannot exist in a vacuum. It requires organization. There has to be some standard on which to base comparison of the competition. There is thus a need for a definition of the scope or basis of what DXing is and a definition of what constitutes a "country." Otherwise the competition has no meaning and no comparative worth.

The establishment that has evolved into the position of giving DXing its form and shape is primarily the ARRL. It accomplishes this through the DX Century Club (DXCC) program. The DXCC Countries List is the standard which is used for not only League awards but for many others as well. The rules for the award and ethics for DXing practices are all part of the DXCC program.

The ARRL also champions the cause for fair play and good sportsmanship in DXing, a necessary chore to help keep the restless troops in line. Articles and editorials appear in *QST* to reiterate the ethics of DXing. In fact, I would like again to refer to the recommended DX operating procedures and guidelines in September 1979 *QST*. This summary of DXing protocol (reprinted earlier in this book) is excellent.

Even though the ARRL (through its DXCC program) is the primary organization for DX activities, it should be pointed out that the other groups, including those which publish the various ham magazines, are also quite supportive of DXing. *CQ* in particular has an award program and all the magazines support DXing with articles and features. DXing is thus a vital and significant part of Amateur Radio. Hams that DX probably constitute the largest single fraction of the total amateur community worldwide. And certainly, in enthusiasm and gusto, DXers make their mark.

Before discussing the DXCC program there are some other awards that should be mentioned. *CQ* magazine has divided the world into 40 geographical DX zones. Working countries in each of the zones requires moderate skill as some contain difficult countries. After confirmation one may apply for the Worked All Zones (WAZ) certificate. Anyone actively pursuing DXCC easily gets the forty zones required for the award. WAZ rules are available for an SASE from *CQ*, 76 North Broadway, Hicksville, NY 11801. *CQ* also has a parallel program for the DXCC countries that also may be pursued.

CQ also sponsors the formidable Five-Band Worked All Zones. Five-Band WAZ is a significant DX challenge. It is a true DX award because the 40 zones cover the entire globe and to work into those areas on five bands is some trick. A ham's Five-Band DXCC, for example, may contain some toughies, but on a band like 80 meters the vast majority of the hundred submitted for the award are the easiest hundred around. Not so for Five-Band WAZ. Since the zones are worldwide, contacts into areas like China or Mongolia are *required*. Achieving this award is difficult and requires considerable DXing skill.

In addition to the *CQ* awards program, there are countless others sponsored by almost every radio organization. There is for example, the Firecracker Award from The Hong Kong Amateur Radio Transmitting Society and also the Diplome d'Excellence de L'Union Francaise from the Reseau des Emetteurs Francais. With titles such as these, it should be obvious that there is something for everyone out there. Most of the awards require some degree of competitiveness and many have continuing levels of advancement (like DXCC) which require moderate to considerable DX skill. Many of the awards center around a particular country and its districts, etc. This gives foreign language buffs a chance to excel in areas that offer other, special interests to them.

Most of the countries' national radio organizations structure awards that require in-depth working of a given nation and/or its territories. For example, the Espana Diploma doesn't just require working the eight Spanish districts but a minimum of 125 EA stations with at least three contacts per district. There is a Worked All Italian Provinces award and what with there being 75 provinces, there is a lot to keep one busy (though actually non-Italian hams can get the award with only 60 provinces worked.) DXers can pursue the German DLD award which offers considerable possibilities. The large Amateur Radio districts are subdivided into DOKs and there are 600 of them. To reach the gold proficiency level one must work 500 of these (a DOK number is on virtually every German QSL card). If you think the Red Baron frustrated Snoopy, try this one. The French award referred to in the preceding paragraph requires working the worldwide network of French territories from Europe to Africa to South America to North America to Ocean and finally to the Austral Continent (e.g., Kerguelen, Crozet, etc.). As you can see, this is both a very French and very DX-oriented award. In pursuing such awards one can't help from getting involved in the culture and geography of a given country.

It should be obvious from all this that there is much to do during the "off time" from searching for new ones, and many of the alternative awards offer parallel DXing activities which simultaneously are important for the DXCC program. A great source for information about these awards is available in a Radio Society of Great Britain publication, *Amateur Radio Awards,* available in the U.S. and Canada through *ham radio* Books, Greenville, NH 03048. This book has complete rules of how and where to apply and pictures of many of the awards so you can pick the pretty ones to seek.

THE BASIC DXCC PROGRAM

The DXCC program is *the* most popular and sought-after award in all of Amateur Radio. Its issuance is carefully monitored by the DXCC Desk at ARRL Headquarters. To obtain DXCC, you must make two-way contact with 100 "countries" on the ARRL DXCC Countries List (available from ARRL for $1.00). QSLs are required for proof of contact. Note that the term "country" for DXCC purposes does not necessarily agree with the dictionary definition. Many bodies of land not having independent status politically are classified as DXCC countries and represent historical as well as Amateur Radio considerations. For example, Hawaii is considered a DXCC country because of its distance from the U.S. mainland. (Land areas of the world become eligible for DXCC country status consideration through application of the specific Countries List Criteria in the DXCC Rules.) There are over 300 countries on the ARRL DXCC Countries List.

The DXCC program is accepted, popular, and works well. It has a lot of tradition behind it and it is supremely worthwhile. The ARRL supports it fully including ancillary services such as the QSL Bureau. DXers have to be fanatic about some DX program and this is it. In addition, supporting DX ends up supporting the ARRL, which is our primary defender on a myriad of issues involving Amateur Radio. The total picture is a healthy, positive one.

APPLYING FOR DXCC

The DXCC award is available to ARRL members in Canada, the U.S. and possessions, and Puerto Rico, and all amateurs in the rest of the world. There are six separate DXCC awards: mixed, phone, CW, RTTY, 160 meters, and satellite (see the DXCC Rules, available from the DXCC Desk, for details).

To qualify for the initial DXCC certificate, you must work at least 100 DXCC countries, receive confirmations in the form of QSLs (which must clearly show your call sign, date, time, frequency and mode), and submit those 100 QSLs, along with the necessary paperwork, to ARRL Headquarters for verification. Once the basic DXCC certificate is issued, the certificate can be endorsed, by sticker, for additional countries (in specific increments) by sending the additional cards to ARRL Headquarters. See the DXCC Rules for details.

When one makes the initial application to the DXCC program

there is frequently some apprehension—wholly unnecessary. For example, it is not true that the DXCC Administrator eats those QSL cards that he doesn't like. It is also not true that cards are subjected to carbon dating and chemical analysis to be certain that something like the Hitler diaries won't slip by. And, of course, the DXCC Desk personnel don't really roll in the aisles laughing at your nervous mistakes on the initial application. In fact, the DXCC Desk bends over backwards to facilitate acceptance of cards and granting of the award. Let us review some of the basics and hopefully dispel some myths.

If you are applying for DXCC for the first time, you will need three things:

☐ Application record for new DXCC (CD-253)
☐ The application form for listing your cards by prefix (CD-164)
☐ A copy of the DXCC Rules.

These application materials are available upon request from the ARRL DXCC Desk, 225 Main Street, Newington, CT 06111. Please include an SASE.

The application record for new DXCC is submitted one time—the first time. Afterward, the DXCC Desk enters your endorsements on the record. This form is similar to the Countries List. On CD-253, countries are printed in alphanumeric order by prefix. What you do is write the suffixes of the call signs next to the countries for which you are claiming credit. The following examples shall explain.

QSL card	Prefix on form	You enter	Country on form
A22GV	A2	2GV	Botswana
MP4BHE	A9	MP4BHE	Bahrain
UK6DAD	K6C/*D*/K	AD	Azerbaijan
YB0ADT	YB	0ADT	Indonesia
DF2LA/5N0	5N	DF2LA/5N0	Nigeria

It's really quite simple; nothing to it really. Just pay attention. Remember that CD-253 becomes your permanent DXCC record. Note the need to underline the D of the UK6 call.

The next form, CD-164, is for the listing of the calls alphabetically with country name, date of QSO, signal report and type of DXCC involved (mixed, phone, etc.). This form is self-explanatory except for the definition of alphabetical. Sometimes a country uses a special prefix and if you put that prefix in strict alphabetical order it may not match the alphabetical listing on the DXCC list. The thing to do is check the main list and insert the country where it is sequentially located on this list. For example, if you put a DM card between DA and DU on your list, it will be in alphabetical order but out of sequence from where East Germany is on the main list as a Y call, between Burma (XZ) and Afghanistan (YA). It is thus helpful to consult the main list but also important to know that cards are not rejected because they are out of the preferred sequence.

One question that newcomers often ask: Suppose I have this second form made out in proper sequence and then some new ones come in the mail—do I have to do it over again if I want these new ones enclosed? No. Add the supplemental calls at the end and they will be checked and counted. Don't add a large number, of course, just those that drift in a little late. (Note: DXer's Murphy's law dictates that just after you've gotten this list in good order, with over 100 countries involved, five new ones come in immediately. Prior to this you hadn't received a new one in several weeks—which is why you went ahead and prepared your DXCC application in the first place!)

The third form is the DXCC Rules. It is not a form in the sense of the application procedure but it is needed as the in-shack reference to the current DXCC status. Keep a copy on hand, as it provides information on the various DXCCs, application procedures for same, how to apply for endorsements, along with the specific criteria on how an entity becomes a DXCC country.

FIVE-BAND DXCC

5BDXCC was created to promote proficiency and versatility in operating on different amateur bands. An enhancement to low-band DXing, 5BDXCC is a good test of the DXers' operating abilities but is well within the reach of all of those willing to work for it. You must work and confirm 100 DXCC countries on each of five amateur bands. All DXCC rules apply (although all QSOs must be made January 1, 1969, and later). 5BDXCC qualifiers receive a handsome certificate and are eligible for a very attractive plaque—just the thing for your radio room wall! (Note: By action of the ARRL Board of Directors, 10-MHz and 24-MHz confirmations are not creditable for ARRL awards.)

The following is a "simplified" version of the current ARRL Countries List.

PREFIX	COUNTRY
A2	Botswana
A3	Tonga
A4	Oman
A5	Bhutan
A6	United Arab Emirates
A7	Qatar
A9	Bahrain
AP	Pakistan
BV	Taiwan
BY	China
C2	Nauru
C3	Andorra
C5	The Gambia
C6	Bahamas
C9	Mozambique
CE	Chile
CE9, KC4, DP, FB8Y, LA, LU-Z, OR4, UA1, UK1, VKØ, VP8, ZL5, ZS1, 3Y, 4K, 8J	Antarctica
CEØA	Easter Is.
CEØX	San Felix
CEØZ	Juan Fernandez
CM, CO, T4	Cuba
CN	Morocco
CP	Bolivia
CR9, XX9	Macao
CT	Portugal
CT2	Azores
CT3	Madeira Is.
CX	Uruguay
D2,3	Angola
D4	Cape Verde
D6	Comoros
DA, DF, DJ, DK, DL	Fed. Rep. of Germany
DU	Philippines
EA	Spain
EA6	Balearic Is.
EA8	Canary Is.
EA9	Ceuta and Melilla

```
EI........................................Rep of Ireland
EL.............................................Liberia
EP................................................Iran
ET.............................................Ethiopia
F.............................................France
FT8W........................................Crozet
FT8X....................................Kerguelen Is.
FT8Z......................Amsterdam & St. Paul Is.
FC,TK........................................Corsica
FG.......................................Guadeloupe
FG, FS....................................Saint Martin
FH...........................................Mayotte
FK....................................New Caledonia
FM.........................................Martinique
FO.....................................Clipperton Is.
FO......................................Fr. Polynesia
FP.....................................St. Pierre & Miquelon
FR.......................................Glorioso Is.
FR......................................Juan de Nova
FR..........................................Reunion
FR..........................................Tromelin
FW..............................Wallis & Futuna Is.
FY....................................French Guiana
G.............................................England
GD......................................Isle of Man
GI.....................................Northern Ireland
GJ, GC........................................Jersey
GM.........................................Scotland
GU, GC.......................................Guernsey
GW............................................Wales
H4, VR4...................................Solomon Is.
HA............................................Hungary
HB........................................Switzerland
HB0.....................................Liechtenstein
HC............................................Ecuador
HC8.....................................Galapagos Isls.
HH..............................................Haiti
HI.....................................Dominican Rep.
HK...........................................Colombia
HK0.......................................Malpelo Is.
HK0.........................San Andres & Providencia
HL, HM........................................Korea
```

```
HP, .............................................Panama
HR.............................................Honduras
HS ............................................ Thailand
HV ............................................. Vatican
HZ, 7Z.....................................Saudi Arabia
I, IT...............................................Italy
IS.............................................Sardinia
J2, FL8......................................Djibouti
J2/A.........................................Abu Ail
J3, VP2G....................................Grenada
J5, CR3..................................Guinea-Bissau
J6, VP2L...................................St. Lucia
J7, VP2D....................................Dominica
J8, VP2S..........................St. Vincent & Dep.
JA-JN, KA......................................Japan
JD, KA1.......................Minami Torishima
JD, KA1.......................................Ogasawara
JT .......................................... Mongolia
JW .......................................... Svalbard
JX..........................................Jan Mayen
JY................................................Jordan
K, W, N, A....................United States of America
KC6.............................................Micronesia
KC6.............................................Rep. of Belau
KG4..........................................Guantanamo Bay
KH1.............................Baker, Howland Islands
KH2, KG6.....................................Guam
KH3, KJ....................................Johnston Is.
KH4, KM.....................................Midway Is.
KH5, KP6.....................................Palmyra
KH5K, KP6.............................Kingman Reef
KH6..........................................Hawaiian Is.
KH7...........................................Kure Is.
KH8, KS6..........................American Samoa
KH9, KW.....................................Wake Is.
KHØ, KG6R, S, T.......................Mariana Is.
KL7 ........................................ Alaska
KP1, KC4...................................Navassa Is.
KP2, KV....................................Virgin Is.
KP4.........................................Puerto Rico
KP5.........................................Desecheo Is.
KX6........................................Marshall Is.
```

```
LA, LB, LF, LG, LJ............................Norway
LU ......................................... Argentina
LX ...................................... Luxembourg
LZ ......................................... Bulgaria
OA...........................................Peru
OD..........................................Lebanon
OE..........................................Austria
OH ......................................... Finland
OH0..........................................Aland Is.
OJ0..........................................Market Is.
OK ...................................... Czechoslovakia
ON........................................... Belgium
OX, XP....................................Greenland
OY..........................................Faroe Is.
OZ ...................................... Denmark
P2...................................Papua New Guinea
PA, PD, PE, PI...........................Netherlands
PJ2, 3, 4, 9............................Neth. Antilles
PJ5, 6, 7, 8.............St. Maarten, Saba, St. Eustatius
PP-PY ........................................ Brazil
PY0...........................Fernando de Noronha
PY0.................St. Peter & St. Paul's Rocks
PY0......................Trindade & Martin Vaz Is.
PZ ...................................... Surinam
S2..........................................Bangladesh
S7..........................................Seychelles
S9, CR5.........................Sao Tome & Principle
SK, SL, SM...................................Sweden
SP ......................................... Poland
ST..........................................Sudan
ST0..........................................Southern Sudan
SU ......................................... Egypt
SV ......................................... Greece
SV ......................................... Crete
SV ......................................... Dodecanese
SV..........................................Mount Athos
T2, VR8....................................Tuvalu
T30, VR1..............West Kiribati (Gilbert and Ocean Is.)
T31, VR1...............Cent. Kiribati (Brit. Phoenix Is.)
T32, VR3...................East Kiribati (Line Is.)
T7..........................................San Marino
TA ......................................... Turkey
```

```
TF..............................................Iceland
TG..........................................Guatemala
TI..........................................Costa Rica
TI9...........................................Cocos I.
TJ...........................................Cameroon
TL...............................Central African Rep.
TN............................................Congo
TR.............................................Gabon
TT.............................................Chad
TU.........................................Ivory Coast
TY.............................................Benin
TZ.............................................Mali
UA, UN, UV, UW, UZ 1,3,4,6....European Russian S.F.S.R.
UA, UN, UV, UW, 1P.................Franz Josef Land
UA, UN, UV, UW, UZ 2.....................Kaliningrad
UA, UN, UV, UW, UZ 9-Ø..............Asiatic R.S.F.S.R.
UB, UT, UY....................................Ukraine
UC....................................White R.S.S.R.
UD..........................................Azerbaijan
UF...........................................Georgia
UG...........................................Armenia
UH...........................................Turkoman
UI.............................................Uzbek
UJ............................................Tadzhik
UL............................................Kazakh
UM...........................................Kirghiz
UO..........................................Moldavia
UP..........................................Lithuania
UQ...........................................Latvia
UR...........................................Estonia
V2, VP2A.........................Antigua & Barbuda
V3, VP1......................................Belize
V4, VP2K............................St. Kitts, Nevis
V8, VS5......................................Brunei
VE, VO, VY1..................................Canada
VE1.........................................Sable Is.
VE1.......................................St. Paul Is.
VK..........................................Australia
VK...................................Lord Howe Is.
VK9.........................................Willis Is.
VK9......................................Christmas Is.
VK9.................................Cocos-Keeling Is.
```

```
ZC...................UK Soverign Base Areas on Cyprus
ZD7.......................................St. Helena
ZD8.....................................Ascension Is.
ZD9..................Tristan da Cunha & Gough Is.
ZF.......................................Cayman Is.
ZK1.....................................So. Cook Is.
ZK2 ............................................. Niue
ZL......................................New Zealand
ZL.......................Auckland Is. & Campbell Is.
ZL.....................................Chatham Is.
ZL.....................................Kermadec Is.
ZM7, ZK3...............................Tokelau Is.
ZP ........................................ Paraquay
ZS1,2,4,5,6........................Rep. of South Africa
H5, S4-S8, T4, V9........Rep. of South Africa, Homelands
ZS2........................Prince Edward & Marion Is.
ZS3......................(Namibia) Southwest Africa
1AØ........................Sov. Mil. Order of Malta
1S.......................................Spratly Is.
3A ...................................... Monaco
3B6,7...........................Agalega & St. Brandon
3B8 ....................................... Mauritius
3B9......................................Rodriguez Is.
3C.....................................Equatorial  Guinea
3CØ..................................... Annobon
3D2...................................Fiji  Is.
3D6.....................................Swaziland
3V ................................... Tunisia
3X......................................Rep. of  Guinea
3Y.......................................Peter I.
3Y.......................................Peter Is.
4S.......................................Sri  Lanka
4U.......................................I.T.U. Geneva
4U.....................................Hq. United Nations
4W .......................................Yemen
4X,  4Z.....................................Israel
5A ............................................ Libya
5B ............................................ Cyprus
5H ........................................ Tanzania
5N ......................................... Nigeria
5R ....................................... Madagascar
5T ........................................ Mauritania
```

```
5U .............................................. Niger
5V .............................................. Togo
5W ..................................... Western  Samoa
5X ........................................... Uganda
5Z ............................................ Kenya
6O,  T5 ....................................... Somali
6W .......................................... Senegal
6Y .......................................... Jamaica
7O ........................ People's Dem. Rep. of Yemen
7P ........................................... Lesotho
7Q ........................................... Malawi
7X .......................................... Algeria
8P ......................................... Barbados
8Q,  VS9 .................................. Maldive Is.
8R ........................................... Guyana
9G ............................................ Ghana
9H ............................................ Malta
9J ........................................... Zambia
9K ........................................... Kuwait
9L ..................................... Sierra  Leone
9M2 ................................... West  Malaysia
9M6,8 ................................. East  Malaysia
9N ............................................ Nepal
9Q ............................................ Zaire
9U .......................................... Burundi
9V ......................................... Singapore
9X .......................................... Rwanda
9Y ................................. Trinidad & Tobago
```

Mailing Them In

The first batch of cards you send off to ARRL Headquarters will arouse feelings similar to those of sacrificing your first-born child. After all, there are probably some *rare* cards included such as 9Y or 8R. What would happen if these were lost? Well, any method of shipping can result in loss of cards and this would, of course, be most unfortunate. Therefore, I recommend spending some bucks to send them the best way to minimize the chances of something going wrong. In my opinion the route should be through the U.S. Postal Service by *registered* mail. This costs more, but there are more checks and balances than with any other system, including the special air express services (which are actually more expensive). Registered mail provides close checking at every point enroute to help minimize loss. Valuable items such as gold coins are shipped this way. That should say something.

When you send it up to the League get a return receipt so you'll know it got there safely. This will save you much anxiety and help to prevent you from screaming at the kids and dog when they greet you in the morning. It's less necessary to pay for a return receipt to the League when they send it back, particularly if you are the very nervous type since you'll be calling often enough to see how your cards are and you will know when they were returned to you.

If a DXCC honcho is at one of the conventions or meetings he will accept your cards personally. Not infrequently there is attendance by DXCC people which is announced on a given convention program. If this is a convention you would like to attend you can bring your cards and forms along and present them in person with trumpets and fanfare. Seriously though, they will take them at ham fests and if you're there it's a good way to transfer your cards.

It has been cited in this chapter that various amateur radio magazines support DX activity. It should not go without mention that several specific DX organizations also do so. The Northern California DX Foundation and the International DX Foundation are two worthy examples. The continued support of organizations like these results in exceptional DX productivity. The number of DX QSL cards with the IDXF stamp on them is amazing. Support of these organizations is most worthwhile.

In addition to this, numerous DX supportive organizations exist in many other nations. The worldwide push for DX activity contributes much to the activation of difficult DX countries. Besides the formal, continual organizations that abound there are also intermittent DX groupings which sponsor specific operations. One should be cautious here since some fly-by-night operations can crop up. Many however are sincere, wholehearted efforts with very effective results such as the Heard Island DX Association's operation by Jim Smith. Amidst much unnecessary controversy Jim's operation came off well and due to his efforts early Heard QSLing was achieved. The other, well done operation at the same time on Heard was also the result of many private contributions, including efforts from the IDXF, Northern California DX Foundation and the Wireless Institute of Australia. The biggest problem on hand now is for those who did not work either of these two operations. Since they were both so effective the "need" is low and those hungry for a contact will have to wait for the next controversial activation. We never said a DXer's life was easy. At any rate, there will come again the time to activate this and other rare DX spots and this will bring out the call for cash. They should be evaluated and if they look good just remember it is ham support that makes some operations possible.

This concludes the chapter on awards which also involves general DX support. The on-going programs around the world which are the result of DXing are substantial. Our involvement with them is inseparable from the sport itself. Our continued backing is thus necessary.

9

The Station

Wholehearted DXing is not done with halfhearted installations. When these poor little people squeak in pileups, their RF is actually evaporated by the larger stations. Why would someone with a low-powered rig and a vertical want to get mixed up in an Abu-Ail or Cocos-Keeling pileup? True, one may eventually get through—perhaps even early—but repeated attempts in truly difficult situations by "half-stations" makes one reminisce of Inspector Clousseau "solving" crimes. To do this over and over, pileup after pileup, should raise questions about a person's mental health. Alas, to get a ham license, one does not need to demonstrate psychiatric competency.

Do not draw the conclusion that I am advocating only the super station to work DX. You know the kind: "We're running the Space Universe 2001 Maxima transceiver here, along with the Steelmasher 5000 linear using a water-cooled 10CX90000A tube; we're putting all this through the Quadrexial Zoom Radiation DX1 computerized antenna system on semi-sequential 100-foot booms at 500 feet. Hope ya hear me OK, little fella." Sound familiar? Well, one doesn't need to go to this extreme either in order to work DX.

A good station *is* essential for repeated DX success. The receiver must be able to "hear" well and be as free as possible from the effects of adjacent stations. The transmitted signal must be strong—a sound transmitter and linear amplifier are essential. The antenna does not have to be the maximum, but it can't be the minimum either. An *above average* antenna is important, first to be able to hear weak signals and second to place the transmitted signal ahead of the majority.

This chapter will describe how to select a DX-competitive station. This seems so simple at first glance, and most hams will quickly conclude that they are quite capable of designing a station configuration for DXing. Yet, it is far from easy, particularly since few of us will have the opportunity to employ maximum, no-compromising quality at each level between the mic and key and the antenna. Compromises inevitably occur, and the knowledge of what compromises can be made and what cannot is vital. DXers go through scores of stations as they learn the subtleties of the myriad of features which make the station effective. And DXers debate endlessly among themselves about station design. These issues are complex and certainly not resolved in product advertising which is obviously biased. And while *some* questions can be answered in product reviews in the amateur magazines, *most* cannot. Product reviews are on the one hand technical and on the other general, so that while part of the evaluation is useful to the DXer, the overall slant is not. A station designed for DXing has very precise requirements so that under the worst conditions, at the worst time, with everything going wrong—a situation in which a routine or regular QSO would not even be attempted—a needed DX contact can still be made.

The chapter will be divided into four sections: receivers, transmitters, linear amplifiers, and antennas. In this age of transceivers it may seem unusual to separate receivers and transmitters, but the receive and transmit chains obviously have independent considerations whether or not they are in the same box. The discussion will not attempt to settle the debate over which rigs or antennas are the best nor will it be highly technical. It will, however, be practical. The text will concentrate on station aspects that you really need, and it will alert you to those that are superfluous.

THE RECEIVER

It is always interesting to listen to arguments over which station component is the most essential. Some propose brute strength and power to clear a frequency so they can subsequently "hear" anything. Others may argue that very fine receivers are the key because they can eliminate or reduce QRM and QRN. And still others concentrate on the antennas so one can copy first and last on band openings and closings and be competitive in between. As has already been stated, in DXing one needs to emphasize, to whatever extent that is feasible, all of these things. Finally, the ancient adage, *you can't work 'em if you can't hear 'em,* is universally recognized.

Given this clear and universally recognized objective, hams then diversify radically on how to *hear* properly. On the air one often observes operators who defend older equipment as being just as good as all that new-fangled stuff. Some may be, depending on how it is maintained, aligned and modified. Much, however, is not. The irony is that with old equipment—*and new as well*—deficiencies in intermodulation distortion, filter leakage, etc., usually go unrecognized. These weaknesses *sound* like QRM or condition noise, so the operator assumes, most incorrectly, that the receiver is OK and the band is lousy. This is unfortunate. It is also unnecessary. Though receiver statistics vary—and manufacturers and testing labs disagree over the specifics—there are some general aspects of receiver design which are uniform and which result in basic standards that produce good results. And yet there is more. In this discussion, receivers will be approached not only from the standpoint of design but also in terms of features and specific SSB versus CW requirements. A good rig can be had if you only plan thoughtfully.

Receiver Design

What is the best receiver today? In my opinion, the following group of *transceivers* could at least be brought forth for consideration of this question: The previously manufactured Drake TR-7A and R7A, the Kenwood TS 930S and TS 940S, the Collins KWM-380, the Signal One CX-11A and Milspec 1030 and the Yaesu FT-One. Certainly these are recognized as some of the most expensive stations. If one has these he's in the top group, right? Well, consider this. The Rohde and Schwarz EK070 receiver (designed by Ulrich Rohde, DJ2LR) is available for amateur use. The *basic* receiver costs $14,000. With options it can reach $20,000. Obviously, there is quite a range of quality out there in the ol' radio market.

I said previously I would not render judgment on which is the best rig (who could?), but I do want to eliminate any prejudice about this controversial section. The reader might wonder what kind of rig I have and inject misinterpretations of what I'm trying to say. If I praise a feature in the Drake TR-7, one will conclude I have this rig and that I'm biased. If I then criticize the same rig, the reader might conclude something else. Let's not make it mysterious. To set the record straight, I have a KWM-380, but I will try to keep specific comments about Collins to a minimum. One logically has the right to take any judgments I do make about the Collins KWM-380 with a grain of salt. I'll do my best to shoot straight, however.

Before getting into the current generation of equipment let us briefly look at the last, for much of this gear is still in use. Oodles of early Drake, Heath, Yaesu, Collins, and other equipment is on the air today. In general, these tube rigs have strengths and weakness in design that have long been well understood. Unfortunately, it is impossible in the scope of this book to compare the advantages and disadvantages of various earlier receivers. In addition, future improvements at the factory level are virtually finished since these rigs are no longer manufactured. Yet, there are literally hundreds of improvement articles in the ham magazines. If you have one of these tube rigs and you want the best performance possible from it, it is worth the effort to research the literature.

Results of such research will be variable. For example, some rigs (such as Drake and Collins) employed permeability-tuned front ends for greatly improved performance. Such a major modification is not easily incorporated in most rigs by the average ham. On the other hand, scores of small but significant improvements on just about every rig have been described in most of our periodicals. Between ham friends, libraries and a letter or two to the particular manufacturer these can be ferreted out without much difficulty. The problem then is making the modifications. If one is capable and comfortable with tube rigs, they can be worked on. If not, and you truly want a better receiver, then stiffen up, get rid of the old, and buy something new.

Ahh—but you say that you have a state-of-the-art tube rig. It doesn't need mods. Bunk! *All* tube rigs straight off the old assembly lines need improvements. For example, let's pick on Collins. In *ham radio* magazine alone, between 1977 and 1981, there were 25 articles for improving the Collins KWM-2 and S-Line. The only conclusion that can be drawn is that Art Collins' engineers didn't solve every problem. The same is true for all tube rigs, and there is a wealth of information out there to spruce and trim up those old babies to be fine performers. And they can compete. Let us focus on one example.

Sherwood Engineering will take your well-designed Drake R4-C and convert it to something better. Much better. They concerned themselves with all the important parameters of reception and addressed them competently. From circuit redesign to additional IF filters, this updated R4-C is very competitive in terms of dynamic range, blocking, noise floor and sensitivity. The company will provide statistics which (suspiciously) place all other rigs under their modified R4-C in one measured category (which will be covered later). In discussing this controversial rating with other manufacturers, one learns that measuring techniques—and opinions thereof—differ widely, and there is no corner on the market in absolute truth in testing. Further, various independent labs draw different conclusions from the same data and from the same type of testing they often get different results. In examining receivers there is, indeed, much fuel for thought and fight. If nothing else is proved, however, Sherwood's Drake R4-C is an example that old tubes never die and don't fade away either; they can, fortunately, be aptly prodded along.

One essential ingredient in using those older stations is maintenance. Aging affects these rigs more than solid-state-equipment, and an active maintenance program is mandatory. Tubes must be checked, bandswitches cleaned and tightened and alignment adjusted. Remember that minor deterioration in these areas is difficult to detect since the band essentially sounds the same. Significant deterioration in delicate DX work will result, however, so diligent maintenance is essential.

Whole books could and have been written on tube rigs and I realize the discussion here falls short of being complete. However, the tube debate is essentially over, since the majority of these rigs for amateur use have been discontinued. I have not tried to rate various designs and features. This has all been done before. What I have tried to do is orient the owner of such equipment to investigate where his rig stands on two fronts: 1) What are the equipment's inadequacies, and 2) Are there feasible remedies available in the literature or from the manufacturer? If, after considering these points, the rig does not measure up for careful DX work, get rid of it.

With the exception of tube finals in some rigs, today's equipment is virtually all solid-state. The improvement of solid-state technology has been so rapid that, unfortunately, early members of this generation are already outdated. Further, the advances are so significant that, for the most part, these stations cannot be easily upgraded. Where inadequate rigs end and acceptable rigs begin is a gray area. It really depends on the manufacturer and when the engineers there made the decision to incorporate the latest developments into their designs. I do not want to get into the controversy of declaring whether a particular group of radios is acceptable. What we shall do here is discuss *receiver design*. If you find that a large number of the considerations mentioned here are not in the design scheme of your transceiver or receiver, the appropriate conclusions can be drawn. Failure not to mention *your* rig in these discussions does not mean the station is unsatisfactory. However, as just stated, if your rig lacks important design features, you should start to worry. (Worry is OK by the way. All DXers worry—a lot. As you study the features of your transceiver and you see things horribly missing, worry freely. It will build character and begin to change your facial expression so that at least you will look like a DXer!)

When you read the advertisements and spec sheets on transceivers, do you notice anything unusual? The tone is almost like stereo advertising. There is talk about dynamic range and intermodulation distortion. All kinds of dB figures are given in terms of noise floor and signal range which sound like promo's for Beethoven and Wagner on digital recordings. What's going on? Plenty.

Previously, manufacturers told you there was a 2.1-kHz filter for SSB and a 500-Hz filter for CW. They tossed out shape factors and then said the front end was great. (Funny, though, if the front end was so great, why did they put in all those RF attenuators?) In short, the ads ignored problems in solid-state technology and many of the early rigs were truly awful. But things have come a long way, baby.

Up-Conversion

Up-conversion is one of the significant developments in receiver design. Prior to this technique, intermediate frequencies were placed within the spectrum of the receiver coverage. Up-conversion places the first IF above the receiver tracking range, thus greatly improving image rejection. An improved image response reduces spurious emissions within the receiver range. It's a neat, simple trick that cancels interference before it happens. It also enables the receiver to provide quality general coverage. While general coverage is not an urgent priority with DXers, it is, nevertheless, an immediate truth that "the ham frequencies are changing." A rig that does not compromise on *any* frequency is superior. All manufacturers that incorporate up-conversion do not so advertise. If it's not clear, look at the block diagram. The first IF will tell you the truth. If it's above the actual receiver range, it's *up*.

Input Filters

High-pass and low-pass input filters are installed to further improve image and direct signal rejection. Ideally, there should be multiple filter sections which compartmentalize the range of the receiver. This improves immunity from strong signals outside of the desired band. Receivers which have only one high-pass and one low-pass filter that cover only the upper and lower ends of the frequency range are inferior. This leaves a wide open "gate" which allows strong signals anywhere in the entire receiver range to enter and cause crossband intermodulation. Generally, you should be able to identify whether or not multiple high-pass and low-pass filters exist in a given rig in the block diagram. This is not found in routine advertising and must be specifically requested in most cases.

Mixers

All the receiver gadgets in the world, such as selective IF filters, variable bandwidth tuning, IF shift, slope control, noise blanker, and the like, are not worth the knobs that control these circuits if they can't handle strong signals. One of two basic mixer configurations is recommended: 1) Quad-arranged field effect transistors; or 2) passive diode mixers. Mixer design is so complex that it is difficult to decipher advertising claims. In general, FET mixers should be of the type that can handle considerable signal power. Some experts suggest that you should avoid weak signal, dual-gate MOSFETs of the 40673 and 3N211 group. Recent evaluations of the sophisticated Plessey mixer conclude that this is probably not the best mixer for high performance receivers. However, one fairly safe general conclusion from the literature can be drawn. Carefully designed high-level doubly balanced passive mixers (that's quite a mouthful) provide outstanding dynamic range and circuit feasibility. Look for them!

Mixer talk is complex and, unfortunately, boring to many.

Mixers are unloved, anonymous creatures that sit in rigs without any panel controls, and might appear to be unimportant to some. Nothing could be farther from the truth. The junk that one hears on the air, and which is assumed to be unavoidable condition noise and QRM on today's bands, is frequently just generated in the receiver. Excellent mixers thus reduce a seemingly invisible but very real problem. In advertising, this problem is rarely discussed and, when it is, the and is frequently misleading. What to do?

Ask! Send for specs on the rigs under consideration and then call and speak to *engineers* or *repair personnel* at each firm. Toss the above mixer names at them and ask for a comparison. These guys are not in sales and usually talk freely. (I've had several admit to cost considerations in terms of rig design.) Also, be aware of the various approaches in promotion. Some firms that boldly talk about mixers are not employing the best circuitry, while others that do not have trumpets blaring have a good receiver design. Don't lose track of this very important feature for a moment. Before you spend good bucks for a new rig, think about how often you or your friends have upgraded because of station dissatisfaction. Very often, sort of unknowingly, part of this dissatisfaction involves the fact that the ham is vaguely aware that someone else is repeatedly copying stations better than he is. The pros know that mixers are important. So should you.

IF Filters

After getting views from several people involved in filter design and manufacturing, it is interesting that IF filters are, in one sense, uncomplicated and, in another sense, controversial. Everybody recognizes now that six-pole or LC filters aren't employed any more. How simple! We all use eight-pole filters of "high quality" and in well thought out design plans, correct? Correct. But who manufacturers these filters? What is their percentage of error in manufacture, what is their percentage of change in aging, what is their percentage of outright failure and, given theoretical maximum shape factors, and what is done in the transceiver design to minimize leakage and insertion loss which dramatically change advertising claims. In addition. what is the difference between mechanical and crystal filters?

Let's tackle the mechanical versus crystal debate first, and make a long story short. There are specialized applications that alter what I'm about to say, but for HF work there is no significant difference between the two types of filters. Several crystal filter manufacturers feel strongly that well-made crystal filters are superior to mechancial filters, but engineers at Collins, who should be partial to their baby, are not so convinced. Crystal filters were chosen for the KWM-380 because mechanical filters will not work in the design scheme; the latter were not rejected because the crystal version was believed superior.

Now, for the tough part. Which rigs use the best crystal filters and best circuit design? Certainly, one cannot look at shape factor claims in advertising and draw any conclusions. Those cute little curves they use to represent shape factors always look so nicely steep and sharp. The truth is that a well-made, eight-pole crystal filter, designed with the same objectives, should be about the same from *any* filter manufacturer. The key is who is the manufacturer and what are the minimum standards and circuit design set by the transceiver manufacturer. Are the filters discrete crystal (the best!), or ceramic or monolithic? Reputations are important here and speak for themselves. A theoretical shape factor in an advertisement is not worth the printer's ink used if it's not backed up by filter and circuit design which insures that *every* rig put out by a given firm will meet certain exacting specifications. Filter leakage, for example, plagues *many* transceivers and deteriorates optimum performance.

Again, call the maker of any rig under consideration for purchase and discuss these details. Ask if they use discrete crystal filters. Ask them what is the percentage of manufacturing error and aging. This should be next to nothing. Inquire about mechanical stability and failure. Ask if anybody has ever tested the rig and found filter leakage. Ask them if their narrow CW filters have any ring in them. Manufacturers who addressed these in their design plans are usually proud of it and will discuss strengths and weaknesses freely. Firms that have cut corners will hedge or make excuses. Back up this investigation by calling the filter manufacturer and double checking what the receiver or transceiver firm has told you. It cannot be stressed enough that splashy ads in the ham magazines have nothing to do with the assembly line, which must first assemble, then align, inspect, and control the degree of failure of the various sections of the rig you are about to buy and use. Think of it as buying a car. That should put it into perspective.

Synthesizers

In a layman's discussion, it is very difficult to keep this subject simple because of, once again, conflicting points of view. It is generally recognized that well-designed synthesizers provide remarkable frequency accuracy, stability, reproducibility and extremely fine tuning. The conflict comes from the following considerations: analog proponents claim low noise factors in VFO circuits, synthesizer proponents claim properly designed synthesizers are as effectively low in phase noise, and to further confuse the picture, a select few manufacturers pursue synthesizer phase noise to costly extremes.

Sherwood Engineering argues that their modified Drake R4-C has the best "close-in" dynamic range available, i.e., signals tested 2 kHz from the desired frequency caused less interference than in other rigs. Analog VFOs don't have the noise problem that synthesizers have, but the gain in frequency accuracy, stability and reliability of the latter is extremely substantial and useful. In addition, as signals move away from the specific close-in range of the R4-C, a border is crossed where well-designed solid-state rigs are far superior in signal immunity. Thus, a strong signal 50 kHz away, which is equally dangerous if it generates noise, will be better rejected if the front-end design considerations previously discussed are observed.

To further confuse the discussion, there is the question of how important the degrees of difference are. The tests which demonstrate the subtleties of variation just mentioned are conducted with very special equipment that often does not equal what happens "on the air." For example, a sideband signal, composed of the human voice and its complex frequency range plus the intermodulation distortion inherent in a transmitted signal, acts very differently in a receiver than the theoretical very pure test tone involved in some of the Sherwood experimentation. In sideband conditions, it is difficult to determine if there is any actual difference between these two receiver concepts. In CW, the variation is more definable but since it takes very strong signals to demonstrate problems, there is question as to which type of immunity—close or more distant—is more useful. In addition, with the employment of newer, *very* narrow CW filters, where selectivity is considerably enhanced relative to earlier narrow CW filters, the actual on-the-air use of the newer solid-state rigs is superior overall, according to most engineers.

Finally, some receiver manufacturers spend more money on circuitry to reduce synthesizer noise. Even so, the difference in the noise figure of ultra-high-priced receivers compared to other well-designed and well-manufactured rigs is only about 6dB out of the over 100 dB of dynamic range involved. I believe at this point we are getting into a sort of technical perfection that may be more than required.

To sort all this out, I simply recommend rigs which employ a synthesizer. The advantages outweigh any possible disadvantage when compared to the alternative analog VFO. Finally, spending a fortune to get every dB of noise out of the way is unnecessary. It's not dissimilar from the high fidelity argument about reducing total harmonic distortion to very small figures. The spec sheet looks nice, but the improvement is inaudible.

Microprocessors

Microprocessors do not really need much discussion. A microprocessor is really an accessory which does legwork for you. A receiver or transceiver can theoretically function without one, but in today's designs, they compute frequency differentials and display the operating frequency, do memory functions, provide scanning and do other miscellaneous chores. Do we need this? Do we need a chauffeur? Do we need a butler? No, . . . but a microprocessor makes the rig nicer. Much nicer.

It might be added that, in one sense, it is not a luxury but somewhat of a necessity. Microprocessors provide the A and B VFO schemes that allow split-frequency operation. While the addition of extra memories beyond this, with touch-pad or panel control, is nice but nonessential, the A and B transceive and split combinations are necessary for DXing. Two VFOs, or an external VFO, can be used, but this is costly and is something else to break down. While microprocessors can malfunction, in general they provide very high reliability.

For your consideration, I have reviewed a number of design factors for current manufacturing of receivers. As stated before, these have been phased in over a period of time as circuit design and improvements were developed. Generally, when looking at current equipment, those rigs in which some, but not all, of these issues were addressed, have a lower cost. Equipment that costs more usually includes most, if not all of the considerations expressed here. Companies claiming to manufacture top-of-the-line equipment should have seriously faced each and every design issue discussed. Some equipment goes to the extreme, representing exact laboratory standards. Given the cost of these units, it appears to me they are for the very wealthy only, since actual on-the-air performance reveals very little difference.

In actually comparing receivers for potential purchase, be wary of published specifications. There are vast differences in the way companies set up tests and, in addition, it is easy to hand pick a hot unit and come up with some great looking figures. Also, attainment of a certain minimum standard may be all that's necessary, even though some firms claim "improved" specs. For example, an intercept point (depending on the way it's measured) of + 15 dBm is excellent. Higher figures, even if they are real, do little to improve actual performance. Specifications may also vary from a maximum used by some firms to minimum, worst-case values, used by others. Generally, reputations and past history are more important than what a fancy ad proclaims.

Because of space requirements or limitations, this discussion is abbreviated. The engineering discussion on these aspects alone demands a whole book. Still, for those who think too much technical detail was espoused here, I urge you to think carefully of the preceding section and the questions it raises. The new and, for many years to come, future generation of rigs will be manufactured under the overall design philosophy discussed. This trend will not be reversed. It thus behooves us to familiarize ourselves with what is going on, if only in a basic way.

I shall next tackle receiver *features*. As a DXer, I can comment on this more directly and critically. For, unlike design, which can be extensively debated, feature arrangement is subject only to what the manufacturers think we want. Since most DXers know exactly what is essential, I can get to the point.

Receiver Features

Receiver features are becoming so complex these days that there is a potentially dangerous element of which the amateur buyer should be aware. This element is know as *knobism*. Knobism is the creeping but steady proliferation of knobs across the front panels of receivers and transceivers. For the most part, knobism is a harmless, deliberately distracting technique brought about by competition in a tight ham market. But it is on the verge of becoming dangerous. There are only a few essential things a receiver has to do, and the *fewer* knobs it takes to do them, the better. Multiple controls that perform one essential function make the quick and efficient execution of that function more difficult to bring the rig

The Yaesu FT-980 HF transceiver can be computer aided.

back to "neutral." (Example: leaving the RIT on accidentally and thus not operating true transceive.)

What does a receiver need to do for us in terms of features? Well, for one thing, it needs some type of multiple selectivity system to tighten it up when necessary. For another, it should be able to eliminate or reduce pulse noise. Also, either an IF shift or passband tuning should be available to "tweak" the signal for maximum hearing when surrounded by QRM or noise. Some type of notch should be available to reduce heterodynes. Receiver incremental tuning (RIT) is an important feature, particularly for CW DXing,

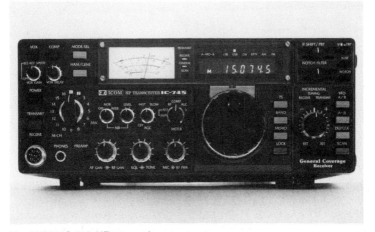

The ICOM IC-745 HF transceiver.

as explained in Chapter 3. Finally, an adjustable automatic gain control (AGC) system should be available to help us hear the desired signal through interference which cannot be eliminated by other means.

That's it. That's all there is to receiving standard SSB and CW signals. In theory, then, we should need ten knobs on a receiver: six for the above described features plus one for the VFO, one each for AF and RF gain and one more to turn it on and off (plus a few for transmit functions in a transceiver). Allowing for all this, have you ever seen a transceiver with about 12 or 13 controls? Let's be more liberal; should a transceiver have 15 knobs? Twenty knobs? Twenty-five? I don't know, but some transceivers are in the forties. Is this necessary or is this knobism?

Selectivity. Without exception, a rig must have multiple stages of selectivity that are independent of transmit functions in the case of transceivers. For example, transceivers which link selectivity to the mode control frequently provide CW wide and narrow positions but don't provide SSB wide and narrow selectivity. If there is no alternate selectivity (such as variable bandwidth tuning), such a rig is operating with compromised performance. In addition, some hams prefer the availability of more than two selectivity positions for CW; however, other active CW DXers find a single choice of 500 Hz acceptable for all applications, even on 160, 80, and 40 meters.

SSB. According to some DXers, tightening beyond the standard 2.1 or 2.4 kHz SSB filter is *essential* for DX work in crowded conditions. There are two basic methods: IF filters and variable-

The Kenwood TS-940S HF transceiver offers slope tuning.

bandwidth tuning. The Drake R7A is an example of a rig that uses a tighter IF filter. Drake employs a 1.8 kHz narrow SSB/RTTY filter for this. The Yaesu FT-One employs variable-bandwidth tuning which moves the skirts of two filters toward each other to provide a range of selectivities down to 300 Hz. Depending on shape factors, a width of about 1.7 kHz is actually all that's needed because the human voice begins to lose its clarity and intelligibility if the filter is sharp and is any tighter.

In testing the differences between these two systems, the IF crystal filter seems to have an advantage. Not one of the rigs available offers both variable-bandwidth tuning (VBT) and narrow SSB IF filters, but most variable-bandwidth tuning rigs do offer narrow CW IF filters. Thus, with all other factors being equal, because the same receiver is being used, switching between the tightest VBT position and a similarly tight CW filter should show which rejects the best. In two different brands, each one's IF filter beat its own VBT system. If this is extrapolated to SSB, I believe the same will hold true. A good, sharp, narrow SSB filter with no leakage will probably be superior to a VBT control which has some type of inherent selectivity loss.

Finally, some reemphasis on the *need* for narrow SSB selectivity. Advertising (surprisingly) doesn't stress this area, and some confusion exists in the amateur community. Very often the uninformed suggest that you can't really "narrow" an SSB signal to the extent necessary to reject adjacent interference without losing signal intelligibility. They point out, correctly, that introducing the increased selectivity to the border where the desired signal is still copyable frequently does not result in a diminution of the strength of adjacent QRM on the S-meter. While it is true that with the tighter filter in line, some elements of really close QRM keep the S-meter high, it is also true that enough is suppressed to bring the weaker, but peaked, desired audio through the adjacent clatter. Both tighter SSB IF filters and VBT can make the difference between a DX QSO being made or not made. Remember, when inserting narrow SSB selectivity, the S-meter evidence of rejection may be subtle with close QRM, but the difference in *hearing* is often dramatic. Insist that any rig you intend to use incorporates some form of SSB tightening.

CW. There is general agreement in both manufacturing and user circles that enhanced CW selectivity is essential for DX work. This recognition is demonstrated in the fact that most rigs today provide a minimum CW narrow position of at least 500 Hz. More

The Rockwell KWM-380 amateur transceiver/LF-HF receiver is probably more frequently identified on the air as "Collins" gear (Collins Radio is now a part of Rockwell).

narrower selectivity is possible and is offered by some manufacturers.

When selectivity gets very tight, two points need to be remembered. First, for selectivity around 300 Hz or narrower, the shape factors should be broadened to prevent ring. Thus, a 200-Hz filter with a three-to-one shape factor will be just as tight in ultimate selectivity as a 300-Hz filter with a two-to-one shape factor *and*, in addition, the 200-Hz filter will have less ring and distortion. A second problem is signal loss in narrow filters. To overcome this difficulty, the receiver circuitry in the CW position should be designed to provide increased gain in order to maintain a constant

The Rockwell HF-2050 VLF-HF receiver with digital signal processing for the coming years.

audio output. When this is done properly, a very nice thing occurs. Signals competing with background noise suddenly become louder relative to this same noise. The selected, boosted CW signal maintains its strength while the equally loud or even louder noise, which cannot now "fit" through the narrower gate, is suppressed. If these two factors—ring distortion as a function of shape factor and gain to overcome signal loss in the CW mode—are dealt with in a given rig, then go for the narrowest selectivity you can find.

A final word about variable-bandwidth tuning is in order. As previously mentioned, tests comparing VBT and the same rig's IF filters tend to show that the IF filter is superior. This is true even when the IF filter is wider than the narrowest position of the VBT. On one rig, a 500-Hz IF filter outperformed its VBT narrow position of 300 Hz. Signal copyability was the criterion, using on-the-air signals in crowded conditions. Combining IF filters with VBT improves VBT performance, however, and in effect makes the ultimate selectivity quite narrow. After looking at the whole picture, I still come down in favor of the IF filter, as long as it's very narrow, combined with very low leakage, and if attention is paid to ring considerations and gain loss. This is all there in the prospectus on a given rig and is easy to decipher.

Besides CW selectivity, another consideration is intermediate selectivity. VBT rigs provide this automatically since selectivity is adjustable. Rigs that only provide an SSB position that functions as a wide CW slot and then offer one narrow position are not ideal. This point is not fatally critical since the final consideration is how effectively tight a CW signal can be narrowed with the idea of making a needed DX contact. Yet, it is nice to "step" down in selectivity for scanning purposes. If a rig offers two or three such steps tighter than its SSB position, then the spectrum is more than sufficiently covered.

Noise Blankers. A noise blanker is a must because frequently impulse noise occurs at the wrong time. There is nothing more frustrating than to lose an important contact because of man-made noise. Today's blankers are fairly sophisticated, and many can take out the longer-spaced pulse of the woodpecker as well. As long as the blanker has provision for variable pulsing widths, I see no real advantage of any one blanker system over another. No one has a "secret edge" on the noise blanker front.

Notch Filter. Another essential feature for DXing, particularly in situations where the DX station is weak, is a notch filter to eliminate heterodynes. Two systems are available: an IF notch or an audio notch. I've heard two engineering points of view here. One claim is that an IF notch filter in a solid-state rig generates noise (not actual direct noise but splatter type noise from adjacent signals which sounds like QRM).

I feel that the IF notch is best and I wonder if some manufacturers aren't pursuing a degree of perfection here that isn't necessary. There is one advantage to the audio notch, however. Typically, an audio notch rejects to a level of over 60 dB, while IF notches cut the interference down about 40 dB. Thus, while the IF notch reduces the blocking effect in the passband, the audio notch is frequently more effective in actually removing the whine of the heterodyne.

Bringing up narrow SSB selectivity again, it should be pointed out that with passband tuning some heterodynes can be eliminated by adjusting the passband within narrow SSB selectivity settings. In really narrow CW positions, however, there is little that can be done since interference in a very tight filter is right on the frequency of the desired signal. Using the notch in this case would also be unhelpful since it would reject the desired signal along with the interfering one.

To wind this up, it is sufficient to say that some type of notch filter is essential for SSB work. Narrow SSB selectivity aids in the struggle as well if passband shifting is available. Finally, if one has really tight CW selectivity the notch is not much use, since this selectivity should exceed the notch's capability in terms of narrowness of the slot.

Passband Shifting. In DX work, some type of system to adjust the passband in order to find peak audio concentration in QRM is a must. As with selectivity there is some differentiation between SSB and CW operation.

SSB. Referring back to the section on selectivity, it should be understood that the ability to narrow an SSB frequency is considered mandatory. Once narrowed, there are further ways to optimize a signal: IF shift, passband tuning and the newer slope tuning offered by Kenwood.

IF shift slants the desired frequency against one side of the filter and rejects interference against this side of the passband. There is a disadvantage in that there is a "favored" side of rejection and an "unfavored" side inherent in this form if IF movement. Full passband tuning has the advantage of having two sides cut with the effect of moving a "window" across the frequency involved to allow constant adjusting to peak the desired signal. Assuming maximum tightness that allows for copyability, passband tuning permits continuous adjustment to find the best spot between upper and lower QRM. Slope tuning on Kenwood equipment allows shifting of the passband (and tightening as well). However, once again, two maneuvers must be made as the high frequency adjustment and then the low frequency adjustment (or vice versa) are tracked together across the IF so as to maintain copyability without losing intelligibility. In conditions that are not too rough, slope tuning allows for good fidelity adjustment, but if sudden interference crucial to a delicate DX QSO arises, I can envision some problem in trying to play these two somewhat competing knobs for proper and critical balance. Nevertheless, once adjusted, it will place the window for optimum signal readability as does passband tuning. One example of several arrangements that I like is the Drake TR-7A. It allows several steps of selectivity to be cranked in with independent passband tuning that is operational at all stages of selectivity in both SSB and CW.

The Signal One Milspec 1030 has synthesized passband tuning. It has to alternate with VFO use and must be activated by a push button. Since this action deactivates the VFO, I see problems in split frequency operations where on receive the operator is constantly tweaking the received signal in the passband and simultaneously shifting transmit frequency for the best spot. This activity is busy enough already and to have to activate and deactivate the passband tuning and VFO is a very unnecessary and time consuming chore. This is all done because Signal One wants a synthesized approach which also allows digital readout of the shifting passband on the VFO LEDs. In this case, it could seriously hinder rapid DXing techniques.

CW. The following explanation for use of passband shifting in CW operation assumes a worst case situation in which the tightest selectivity possible has been engaged. Once again, the idea is to optimize the signal, but in a narrow CW slot, a shift in passband loses the signal. Though a limited amount of "tweaking" can be done, the concept here is to move the passband with a simultaneous frequency shift to select a beat note as far away as possible from QRM. This technique is particularly useful in split operations in which, for example, the pile-up is up band with the edge of close signals spilling over onto the DX station's frequency. By moving the frequency and tight passband away from this noisy fringe, copy is improved.

As with SSB, passband tuning is superior to IF shift in accomplishing this in CW operation. This is because passband tuning moves a true two-sided cut window across the desired signal's frequency to find the best spot for copyability against QRM on the adjacent upper and lower sides. Both the Kenwood TS-930S and TS-940S have a very nice feature in their CW pitch control. It moves the passband while raising or lowering the beat note at the same time. Assuming simultaneous tracking is good, this allows for quick shifting without having to change the VFO. Of course, one must always remember that QRM which lands right on the DX station's frequency is virtually unremovable, even when using very narrow CW selectivity.

In summary, passband adjustment is another essential element in careful DX work. Passband tuning, with its true, two-sided slice above and below the desired signal, is superior to the limited IF shift in accomplishing its mission. The only current variation on this theme is the Kenwood TS-930S slope tuning. This enables the operator to move the same passband window across the operating frequency yet requires sequential adjustment of two controls. However, regardless of the system, peaking signals with some type of passband adjustment will occasionally be the difference between a successful DX contact and an unsuccessful one.

AGC. Ahh, what a sleeper! I can't recall any significant advertising stressing the importance of AGC (automatic gain control) circuits. Even serious DXers, searching carefully for every advantage, frequently fail to appreciate the importance of flexible AGC systems. January 1979 *QST* addressed this as follows: *It seems as though most receiver engineers consider AGC a necessary evil, so they take the "let's hurry up and get it done" attitude.*

The point here is that excessive AGC, espoused by many manufacturers, results in a flat response, lacking depth and presence. Others favor a PIN-diode attenuator to provide a *linear, variable-slope AGC system.* This is incorporated in just about all of the serious receiving designs of today, though it may be necessary to ferret this information out of any engineering department.

While I, too, favor this approach, which provides for depth of sound to avoid listening fatigue. I also wish to stress the need for a special role of front panel AGC selectivity: the ability to turn it *off.* In DXing one needs to hear and copy regardless of how good the signal sounds. When any type of AGC delay is employed—fast or slow—there is a receiver pause, no matter how quick, that can mask a hard-to-copy signal in the presence of interference. By being able to turn selectable AGC completely off, the listener employs an instantaneous recover time from serious QRM and QRN which allows for significant copy improvement in some situations. Though such a signal hardly sounds beautiful, the ability to copy required DX information is often remarkable. The recovery time when the AGC is off is strictly instantaneous, thus the desired signal competes with noise and interference on a real-time basis which allows copy in very bad interference circumstances. Of course, audio circuitry in the receiver must be well designed to prevent excess distortion in which a weak signal is competing with louder QRM and QRN. Thus, excellent audio handling characteristics are necessary in the receiver.

To briefly recap, selectable AGC with an off position is desirable and on certain occasions even necessary. The instantaneous recover time in high interference conditions sometimes allows a needed signal report or other information to be brought through the wall of noisy QRM and QRN.

This concludes the section on receivers. I have covered receiver design and features. While much has been said, much has not, and it is imperative that the buyer beware. The requirements for DXing are different from many other low-key operating styles. The equipment must reflect this difference, not just in what's represented on the flashy front panel, but also in what constitutes the inside: the unseen, more hidden and secretive "guts" of the receiver. This has not been a definitive treatise on receivers. That would require a complete text in itself. And yet, it should be more than ample for the discerning amateur to assemble an excellent station. With good common sense, you will be able to get the right rig with the right combination of design concepts.

Fortunately, the inevitable push for quality is forcing design considerations in the right direction, and more manufacturers are now putting out improved equipment. This does not mean that every rig is great—or even good. Diligence in purchasing is clearly necessary. In addition to smarts, however, it is also helpful to have a few bucks handy.

THE TRANSMITTER

Luckily for you and me, this section is not going to be nearly as long as the one on receivers. While it is true that transmitter design is complex, it is also true that, technically speaking, the design concepts are relatively unimportant in a book about DXing. Transmitters today are almost universally "clean" in terms of transmit quality. Almost without exception, when a bad sounding

Behind this "innocent" panel lurks a pair of 4-1000 tubes. This amplifier is effective in subtle pile-ups.

station is heard on the air, it is due to a malfunction of equipment or RF feedback into solid-state components such as speech processors. These are not design faults. These are particular station deficiencies. Nevertheless, there are some transmitting considerations that are important to DXers.

Final Amplifiers

When buying a rig, a decision regarding tube finals versus no-tune solid-state equipment must be made. Another difficulty is resolving the problem of transmitter output as a function of the minimum driving requirements of various linears. These quandaries should not exist, but manufacturers continue to provide the dilemma of making such choices. It would seem to some that low-powered tube finals are destined to soon be nostalgic memories. However, Yaesu, for example, balanced its entry of the all-solid-state FT-One with its tube-final FT-102. Best to cover all bets is the manufacturing mentality here.

There is really no mystery in the decision-making process. If I had a Drake TR-4 driving a Drake L-4B, for example, I would be hesitant about switching to a solid-state rig unless I was ready to switch linears—say to an Alpha that requires lower driving power. In making my plans, I would consider two things: one, as just stated, solid-state finals are here to stay and, two, linears designed with lower drive requirements put out just as well as hard-to-drive ones. In making any station decisions, this combined thinking about driver and linear are all that's necessary. Simply think long term. All station components get changed—even linears, especially with the WARC bands—and buying a new rig with tube finals to excite an old hard-to-drive linear may well be a short term and short-sighted idea. At any rate, I believe we are looking at the last generation of tube finals in HF transmitters.

DXers are frequently contesters, and it goes without saying that the no-tune solid-state finals allow for rapid frequency change. On a somewhat less obvious scale, a beginning DXer often encounters multiple pileups for countries which are not rare but are still not necessarily easy. This is particularly true when working between the 150-and-250 country levels. An example might be your first TU, XT, or OX. Suppose all these countries are on at the same time on different frequencies and bands. Solid-state rigs with memories for quick frequency changing are very nice, I know. When I was doing this, I didn't have the capability. Now that I do, one new country comes along every few months. To sum up, I would not pur-

Homebrew is alive and well. A handsome assembly for a pair of 3-500Z final amplifier tubes provides this DX station a significant signal.

chase a new transmitter with tube finals. The benefits of rapid, no-tune frequency shifting and the potential for memory recall are too important to pass up even if it means one has to purchase a new linear. (Not infrequently, some of the solid-state rigs drive a linear like an L4-B quite well, if not "all the way." Some solid-state rigs—even within one brand or model—put out more than others, and that old linear may do quite well.)

CW Features

In looking at CW requirements for transmitters, remember that the majority of DXers are not *primarily* CW operators. CW buffs who want to raise flags have their justifiable pride but one only has to look at the totals for Mixed, Phone and CW DXCC to see that the majority of DXers do not concentrate on CW. If one is a dedicated CW buff, one does need to look at particular CW rig features; however, if one is a general operator specializing in DX, then CW features need not dictate final purchasing policy. Even so, almost all serious DXers have considerable CW capability. And for this purpose certain CW features are nice. Separate VOX delays for SSB and CW enhance switching modes. Frequency zero-beating by a variety of methods enables one to be on frequency accurately. This is very important in CW transceive operation and something which is frequently misjudged when the transmit frequency is automatically offset in the CW mode. There are several ways to ensure that one is on the correct frequency. Rig advertisements or spec sheets usually outline manufacturing approaches to these considerations and need to be judged accordingly.

Speech Processing

I am an advocate of selecting a rig that has some form of internal speech processing. Another advantage to internal processing is in general use when not DXing. Since internal processing generally sounds better (regardless of the type of circuit), it usually may be left on constantly. I have found just about all manufacturers who provide internal processors offer both effective and good quality circuits. And fortunately, most manufacturers now offer this important option. Remember that a processor is to a DXer what a microphone is to Howard Cosell, or to a politician for that matter.

There are a few other transmitter considerations from the standpoint of DXing, such as certain mobile applications and appropriate power supplies, line voltage variables for DXpeditions and other unique installations, such as in aircraft or boats. In general, however, this book is about *working* DX and not about *being* DX. As such, I need not delve into these other specifics in these pages.

If I could make one generalization, I believe—so far as *transceivers* are concerned—that if one satisfies DXing requirements for a receiver, one will be satisfied with the accompanying transmitter.

THE LINEAR AMPLIFIER

First it should be quite clear that I'm not going to compare an amplifier running a 4CX1000A in Class A to one running three 8877s in Class B or to one running 10 3-500s in Class C. What is important is making sure that your exciter will provide the proper drive so that the amplifier will operate at its rated power with a minimum of spurious emissions. I recommend finding people using the linear you are considering, and if necessary, hooking up your rig to it. At the very least, get all the specs from the factory *and* talk to the engineering department there. Finally, get on the air and start ragchewing about linears. If you spend enough time, you'll find out how much drive any linear really takes. Remember also to be wary of advertising.

When shopping for linear amplifiers, you pretty much get what you pay for. Quite simply, rugged components, heavy duty power supplies, adequate cooling, etc., raise costs, and the price of a linear amplifier pretty much reflects its capabilities whether commercial or home brew. Incidentally, high cost shouldn't deter a DXer from getting whatever linear amplifier he or she wants. There are various convenient methods for funding such projects, such as second mortgages, selling family jewels, selling family members and the like.

I would like to reemphasize one main point: every dB is important in DXing, every single one. I hear people shrug off power differences with statements to the effect that once you've cleared a thousand watts, going to 1500 won't make much difference. Theory be damned here; I've heard too many on-the-air battles when I knew what the stakes were and what amplifiers were involved. The results verify what I am talking about. This does not mean you cannot work much DX without a massive linear. Of course you can. But depending on how much you compromise on power, the challenge gets more formidable.

ANTENNAS

Nothing can get two hams arguing quicker than to bring up the question of antennas; because there are so many personal experiences with antennas, a large variety of strongly held opinions emerge, quickly. And, until the next experience comes along, temporary ideas get stuck in fast setting concrete. But even experience does not give one ultimate wisdom. After years of debate, experimentation and additional on the air use, the "experts" (whoever they are) are still divided. Cass, WA6AUD, continually referred to the *mysteries* of DXing. Well, in terms of mysteries, antennas have no equals.

A recent article in *ham radio* reminds us of the dilemma. This article reexamined the classic quad-yagi debate under the title, "A Quad Owner Switches." A close examination of the text reveals conflicts in the results (acknowledged by the authors) and, finally, conflicts in once again trying to definitively answer this simple question: which one of these two antennas is best.

Far be it from me to attempt to finally answer this wonderful

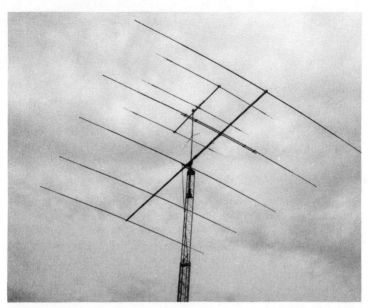

Here's five elements for 20 meters, with a three-element duobander for 10 and 15 meters on top.

question lest some yagi and/or quad owner come after me with elements to lance my heart or wire to choke me with. I can do something in this book about this sensitive dilemma, however. First, I have reviewed a lot of the literature and I can give a summary of opinions. Second, from published data and collective experiences and viewpoints some conclusions seem to emerge. Although what I'm about to write will never satisfy everyone, I bet a lot of conclusions brought forth will draw grudging acknowledgment from the antenna community. If it seems ho-hum to anyone, then you're not an antenna fanatic, a condition that may or may not be a healthy one.

Before we get started, let us define our scope of interest. Hams usually use various dipole configurations to solve their 160-meter and 80-meter antenna problems. The problems here are not too substantial from the standpoint of debate and as such will not be covered. Forty meters presents more of a choice of possibilities but, once again, decisions are obvious and are only subject to space and cost limitations. I'm not shortchanging anyone in the discussion at this point, because despite the low-band buff's enthusiasm, general DXing does not require an exotic, controversial low-frequency antenna. (As for 10 MHz, this band will require years of use to formulate DX patterns and antenna solutions, but I have no doubt that

Aluminum sky! Stacked monobanders for the serious signal.

this will eventually be a great area for a variety of antenna concepts to be developed. Judging by the lack of concensus on various other antenna "dogmas," however, it might prove to be a long wait for any definitive direction. Nevertheless, as the sunspot cycle fades and 10 MHz becomes an important band, the experimentation will proceed in earnest. I can't wait to see the first 10-MHz four-element quad at 100 feet.)

What antenna areas am I going to review? Obviously, I shall cover the quad-yagi debate above 10 MHz. This, however, is

A two-element triband quad that is relatively small but effective.

neither a small nor unimportant debate. The majority of DX pursuit takes place above 10 MHz and with these two types of antennas. The higher frequencies are where the greatest pileups and most difficult chases occur. With these come the greatest conflicts and the greatest challenges from the standpoint of competitive DXing. And it is perhaps no coincidence that the great quad-yagi antenna controversy centers in the bands inclusive of 14 and 28 MHz. For it is in this spectrum that antenna size permits the most experimentation—and hence the most quarrels.

If I had a choice of having any antenna system, I suppose I might select something like the one on the cover of June 1980 *QST*. This type of antenna is literally a steel monster. It consists of a four-legged microwave tower 126 feet tall, topped with an 80-foot Sky Needle. The Sky Needle is set into the tower with its base at 106 feet thus the total height is a mere 186 feet. Though not shown in its final form the completed antenna is designed to have three stacked *pairs* of 6-element yagis. Total: 36 elements. Each of the three booms that hold the pair of antennas *alone* weighs 400 pounds. Computers sample the received signal for maximum strength across eight combinations of antenna "bays" every few milliseconds and vacuum relays connect up the best choice at the time one goes to transmit. This antenna could probably provide accommodations for several thousand birds. And it's only for 20 meters!

If you recall, I said at the beginning of this book that one usually does not have the best at all levels of a ham station from mic and key to the antenna. A station usually represents compromise. This antenna demonstrates this very well. And yet, for a DXer, this "maxi" job may be a bit much, since this antenna probably requires considerable on-going maintenance and work. It is probably a toy for the antenna buff rather than an active DXer who should be *on* the air, not *in* in the upper "bays" where the crows live.

I shall thus confine the discussion to conventional yagis and quads. For the purposes of DXing interests, I have divided antennas into three general groupings: large, medium and small. Simple enough. Large includes big monoband wide-spaced yagis and big triband quads. Medium incorporates large triband yagis, three-element monobanders and intermediate-sized triband quads. Small covers "regular" three-element triband yagis and small triband quads. Bear in mind that this is a general comparison, not a construction article. I shall compare quads in general with beams in general and not try to sort out the intricacies of different designs which would be the legitimate subject of an antenna book. Now, to the categories.

Large

In this group, there are the wide-spaced monoband beams consisting of five-plus elements on each band. The quad in this group is usually four elements at close-to-optimum spacing. I am excluding five-element or larger quads, because these babies, no matter what theoretical gain they may have, are simply a lot of trouble to keep on the air, not to mention *in* the air.

Before comparing these two antenna types, the following words from Orr and Cowan's book, *All About Cubical Quad Antennas,* are worth quoting at length because a single antenna article in a magazine, excitedly researched by an enthusiastic antenna owner, will frequently try to draw some hard conclusions. The following words are indeed wisdom when it comes to judging published data and claims.

A direct comparison between the effectiveness of a quad and that of a yagi is difficult to make and the results of on-the-air checks between the two types of antennas are often inaccurate and confusing. Laboratory field strength measurements of the power gain over a dipole of either type of beam antenna may lead to results that are controversial and open to various forms of interpretation. Unless such tests are run with extreme care on a well calibrated antenna range, the difference in power gain between two antennas of approximately similar size will be lost in the inherent measurement error of the test set-up. Antenna tests repeatable to an accuracy of a decibel or better are difficult to manage even under the best of circumstances.

Try to think of this the next time you read a high-powered antenna article. Now, from my research, the consensus from a variety of sources—too varied to pin down to specific references—is that *very probably* large yagis are better than large quads, even if only slightly. This assumes that the yagis are up high. A quad will outperform a comparable yagi at lower elevations and if there is an installation limitation a quad may be one's best antenna. What I have said will ruffle many feathers (keep in mind I am a *quad* owner) and come as no surprise to those with other prejudices, but this is the general conclusion which comes up in trying to "average out" the myriad of opinions. Since both antenna systems are within the reach of most serious DXers, I would, after much considera-

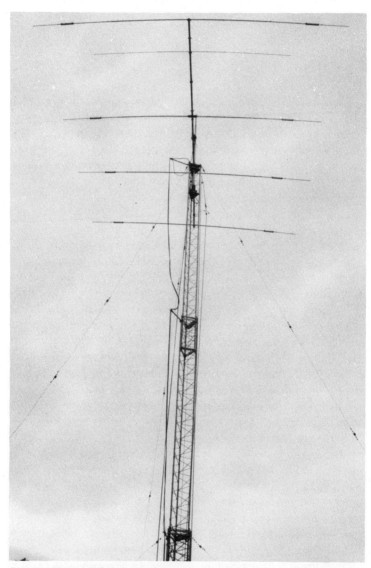

Clean, neat, and effective. This is not an exotic antenna, but that's the point. The man behind this antenna, Bob Robertson, W5OB, has 313 countries.

tion, recommend to those without space or financial limitations that they choose the large, well-elevated yagi antenna installation. These conclusions may seem less dramatic to the non-participant in antenna controversy or to those who already "know" what "best" is, but I assure you that this is still an on-going and hotly debated subject in many circles.

In the years to come, someone may be able to accurately and scientifically resolve the quad-yagi question but I assure you that at this moment this has not been done. Perhaps Overbeck and Bell's opening remark in their antenna article in *ham radio* (May 1982) best summarizes the enigma: "Thousands of words have been published about the relative merits of cubical quads and yagi antennas, and probably millions of hours have been spent in on-the-air debates of the matter." In the near future only the numbers just quoted will increase, not definitive answers.

Medium

For those hams who don't have the capability for large antennas, let us now turn to the next category. This group includes large tribanders (e.g., the Hy-Gain TH7DX or TH6DXX), three- or four-element monobanders (sort of rare now) and the three-element triband quad. It is here, in this middle category, that I believe Bill Orr's theories about a quad's strengths begin to come to fruition. Quad zealots state that, element for element, they can do better with a quad than with a yagi. This seems to hold true. A three-element quad will usually outperform a three-element monoband system and will equal or beat a large, multi-element tri-band yagi. At this level, the quad emerges as a slight leader and, except for weather limitations, I endorse it as the favorite in the medium category.

From the standpoint of construction, it is also a moderately easy antenna to assemble and install. Even though a four-element quad is not much bigger than its three-element little brother, the four-element version crosses a border and becomes a more difficult antenna to install and maintain. For its size and effectiveness, the three-element version is much easier to put up and keep up.

One other point here is that even though the yagi systems in the medium category are smaller, considerable tower height is still necessary; in fact it's even more important *because* the antennas are smaller. Thus, assuming one has chosen the medium category because of cost and space limitations, with a quad system one can employ a lower cost, lower elevation tower without compromising much on effectiveness. The medium quad becomes much more effective when compared to a medium yagi if both are at a height of 50 feet, for example.

Though I wanted to keep the antenna section as simple as possible and discuss quads and yagis very generally, there is one antenna that needs some additional comment. The KLM KT34XA tribander is physically in the medium category, but its performance seems to be somewhat better. Its spacing makes 10 meters its top band, with decreasing performance as it goes lower in frequency. Overall, it keeps a close watch over some systems in the large category. Though bigger than other tribanders in the medium group, it is definitely smaller than the antenna systems in the large group. It's a very good performer for its size.

Small

Antennas in this category include the "regular tribander" three-element yagi, usually with a boom of around 13 feet and occasionally with a boom (such as the Mosley Classic 33) of 18 feet and the small, two-element quad. You simply can't beat the quad here. A two-element quad will run circles around the average three-element triband yagi and will compete with yagis in the medium category as well.

In addition, these small quads are lightweight, easily supported on light duty towers, easily rotated by TV antenna rotators if necessary, easily repaired if the wires do break (use stranded wire and this is rare), easily tuned since they are so simple and, finally, easily purchased (they tend to be cheap). Because of the box configuration of the quad, it does require extra clearance of roofs and other adjacent objects, but it also has a smaller turning radius and thus fits between tall trees and remains within narrow property lines better than a yagi.

The final consideration, however, is that the two-element quad is probably the meanest dB gainer for its size around. At modest heights (50-60 feet), it keeps company well with bigger arrays.

YOUR STATION

Amidst all this heady talk of rigs and antennas, it must be acknowledged that virtually all of us are limited by the all-too-real constraints of location, local ordinances, and finances. Tremendous DX success can be yours, though, even from a typical city or suburban lot. You can achieve it with a tribander or quad at a reasonable height, some wire antennas, a good transceiver, and an amplifier. You do not need some sort of death-ray aerial to have a good signal in today's Amateur Radio. And remember that skillful operating can add many dB to your signal. This is true DX Power; not power measured in watts, but the power of a positive mental attitude (along the lines suggested in Chapter 3). The power comes in the form of the creative urge for self-improvement and the healthy desire to improve one's "station in life." Equipment specifics aside, remember that the objective is to maximize your signal quality and station effectiveness within your own particular considerations. And then get on the air and have some fun!

10

Future

As time marches on, a truly astounding technological revolution will occur. The hint of it today is already impressive. These developments will dramatically alter total communication techniques; in turn, these developments will cause a restructuring of Amateur Radio. It thus follows that DXing will unavoidably be altered as well.

For the most part, imminent changes will not be all that drastic; however, as we cross the century mark, things will begin to race forward. Possibly in the first or second decade after the year 2000 some or all of what will be discussed here might have come to pass. Much of predicting is, by nature, speculation. But the current developments in the various electronic fields add considerable logic to the chore. It is thus safe to assume that these changes are not only predictable but unavoidable as well.

Change is not always for the best. For example, though I love jet travel, I miss the grand old trains that I remember as a child. And though steel and glass skyscrapers are nice, a visit to an old European village with cobblestone streets reminds one that the past is often more beautiful than the present. And so it may be with Amateur Radio. There is the real possibility that further automation will make ham radio a push-button activity. DXing will clearly be affected but it is hoped that there will be some—perhaps many— areas of the world that will not change too fast and the sport will remain just that: a true sport. In this chapter we shall look at the and what it might hold for us, both as general hams and specifically as DXers. Ironically, some of the changes that will occur will not be directly related to Amateur Radio. They will instead affect other communication areas and the population in general and, even more ironically, we will get band expansion as a result. That's right, I mean even *more* frequencies in the future will be coming our way.

While this at first seems to be grand news, and it is in many ways, the technology that is causing it in other areas will simultaneously be affecting ham radio. Thus the expansion of frequencies will coincide with developments that will greatly alter the way we find and have QSOs with other stations around the world. Just how much fun such QSOs and DXing will be in the next century remains to be seen. I have hope, as shall be shortly discussed, that technology will not smother the sport and make talking to Singapore as reliable as the telephone—well, as reliable as the telephone is *sometimes* in that part of the world.

In this chapter, both the short term changes and long term expectations will be covered. Obviously, not too much is going to happen on the short haul, but the seeds are being sewn which will grow and produce the more dramatic effects that are to come.

THE IMMEDIATE FUTURE

From the standpoint of equipment the technological emergence of solid-state circuitry from the primitive to the modern world has essentially been accomplished. There will be some improvements—and there will always be more knobs—but the essential functions and state of development of rigs today will be about the same for some time to come. With the exception of the various RTTY modes, which are undergoing dramatic development right now, the various other modes generally employed will remain the same. Hence, the basic style of Amateur Radio will remain unchanged and the level of rig development will not be altered significantly in the immediate future.

There will also be no significant band additions in the immediate future. What we will have is, for the most part, what we got. If some of the developments talked about in the upcoming sections of this chapter were to occur sooner additional and expansion is possible. But for at least the next decade and probably for the one after significant band expansion is unlikely.

Thus, for all practical purposes, amateur communication, as we know it now, will require the same general type of equipment, as we know it now. The general style of QSOing, and searching,

QRM dodging and propagation seeking will also go on—for awhile.

DXing will also not change too drastically over the next decade or two. There will be the same elements of the hunt and combat and the same frustration over QSOs missed and QSLs that are late. There will be rare countries and most wanted lists for some time to come. DXpeditions will continue to occur, putting on normally inactive lands.

DX talk will not certainly change for considerable time. There will be the almost eternal debates, fights, advice, help, disgust, joy and enthusiasm that propel DXers into animated conversation everywhere. This is our milieu and it shall not be abandoned. Our right to agree and disagree will each other and others is engraved in stone. Such are the prerogatives of the Deserving.

The DXCC program itself will continue to reflect world realities. Politics and war will alter the earth's surface as it always has and countries will change, evolve and merge. And inevitably, the DXCC list will change accordingly. Wise words introduce the Peter Pan story to the effect that all this has happened before and all this will happen again. History has not learned to stop repeating itself.

One rewarding thing about DXing is that after a somewhat recent war—the Falklands'—and after a relatively major crisis—Poland's—the amateurs of Britain, Argentina, Poland, Russia and the U.S. (the main antagonists, one way or another) have continued to speak with each other on the air. Many other serious problem areas exist and Amateur Radio is absent from too many additional countries. It is a sad absence. Like the symbolic Ping-Pong match that brought China and the United States together, the mere existence of Amateur Radio in certain countries could bridge political gaps in an unprecedented manner.

As radio amateurs, with international experience, we've always had that special feeling that, regardless of politics, we will talk with and/or help anyone. It is that special international atmosphere of Amateur Radio that allows us to be steadfastly loyal to our country and its interests and yet simultaneously allows us human contact with those nations with which we are politically opposed. DXers are inherently involved in such matters and are usually cognizant of world developments. It should be hoped that the future will allow Amateur Radio to play a natural role of easing tensions. But it probably will not happen that easily. And yet, worldwide, we are all here, ready to be tapped.

Such problems aside, we will see other developments in the

immediate future. Satellite communication will increase in vibrant programs supported by Amateur Radio worldwide. The achievements in this direction will be quite significant and DXing via satellite is a natural evolution of DXing in general. Even nicer is the gentlemanly approach to same, something absent from 20 meters. And yet, exciting as this is, aggressive satellite DXing will be limited. Deep involvement in pursuing high country totals will still require extensive HF work. Such is the immediate future.

THE DISTANT FUTURE—FUTURE PERFECT?

The long-term changes that are likely to occur are the results of electronic developments that are underway now. For example, for those of us who are stereo buffs, the digital world is here now and it doesn't take a genius to see its ultimate extension into amateur communication (already here in packet radio but with tremendous potential and further impact ahead). Before that is addressed, however, consider another development that will allow for one of the major developments in Amateur Radio: band expansion.

Technological Changes of the Future

Several developments will be going on over the next decades which are going to dramatically affect communication of the populace in general. Three general things will happen: 1.) Satellite communication will increase; 2.) Alternative methods of transmission, such as those utilizing fiberoptics, will be extensively developed; and 3.), Integration of these technological advances into home use will render traditional communication unnecessary. In the case of satellites one has only to read the newspaper to realize the the potential of the space shuttle. It won't be too long before a considerable number of satellites are responsible for massive amounts of long distance communications. This is not a frontier or research area. It is simply a matter of getting them up there, though certain new communication modes may be employed. This almost total switch to satellite communication will have profound effects on communication in general.

Alternative means of traditional communications are already highly visible in cable operation which, if totally implemented, would result in the discontinuance of traditionally transmitted television. Even before we switch to fiberoptic communication, cable will expand to provide a variety of two-way services. Ultimately, however, all services—television, radio, and computer—will be transmitted via systems such as those with fiberoptics. This change, when fully implemented, will further reduce traditional forms of on-the-air communication.

Parallel to the developments described in the preceding paragraph will be the evolution of the home as the primary unit or place of both being a dominant entertainment center and of doing family business. It will become commonplace to accept this total input/output barrage and, as everything is phased in, the home—not industry—will put regular TV and radio off the air. It will, in the end, be a virtual revolution in what the "home" actually is.

These developments are not really probable—they are virtually inevitable. Their combined impact is remarkable considering the effect it might have on what we now regard as valuable frequency allocations. Much of the current high use spectrum of frequencies might become commercially useless. If commercial use (and government-military as well) becomes valueless, these frequencies will be turned over for "hobby" utilization. If this occurs, it will impact Amateur Radio.

The reason for the switch will, of course, be the reliability and efficiency of the newer communication systems. The high efficiency of newer systems to handle a multiplicity of signals and functions into and out of the home, for example, will make even local on-the-air transmission unnecessary.

While all this is going on, there will be simultaneous development of newer generations of computers and further digital progress. The natural alliance of ham radio with these technological developments will stimulate considerable change in amateur equipment and actual radio technique. The rig will be bound to these newer computers and as a result frequency selection and digital communication will alter our whole manner of having a QSO. Early variations of this are already in the operation with such systems as bulletin boards in which RTTY stations trigger a response of another station through computer activation—-the ham on the other end doesn't have to be home.

Suffice it to say that ham rigs probably will be continuously switching frequencies according to computer data. Digital coding on both ends of a QSO allows two or more amateur stations to track each other as best frequency and propagation are calculated. QSOs will be literally scattered across the bands in bits as these frequency and propagation factors are sought out and utilized. Codes for the equivalent of a general CQ or a call for a specific country will be employed to generally or selectively initiate a QSO. Corresponding coding on the other end alerts the operator to such a search . (One interesting concept is possible instant language decoding; communication initiated in one's own language is translated by computer into the person's language with whom he is speaking—and vice versa.)

The effect of all this will be to greatly enhance the ease and reliability of both initiating and maintaining a QSO. It will, in a very real sense, have to be more of a push button operation. Traditional seeking (tuning) across the bands with knowledge of propagation tricks and where DX hangs out will not be required. A lot of the art will be lost and, in addition, we won't even have the satisfaction of being reasonable electronic tinkers. The stuff will be far too complicated to play with and, as a result, for the average ham it will be much like using the telephone. That simple tone pad, we well know, can activate billions of dollars of equipment including satellites without us giving so much as a thought to the electronics involved. It should be very easy to see that, in this atmosphere, the very meaning of an Amateur Radio license could be completely redefined.

How's DX?

Well, in a word, uncertain. DXing will be changed significantly and much of the change will be for the good—advancement in technology always is—and yet some traditional aspects will be lacking. Future DXers, including us then older folks with them, will be doing whatever the sport calls for, but some aspects, which are competitive, will be absent. Going back to sailing for a moment, it is interesting to see this sport stronger than ever. Sailing is primitive (though with energy costs changing this may not always be so) and yet its adherents cling to it for reasons that are quite clear to those of us involved. Sailboat racing is hardly high speed—in the face of high-powered, engine-propelled boats—but its subtleties and challenges lure us back to it again and again. Even here, sailing is progressing with speeds in certain boat types that are thrilling—*especially* since it is without engine propulsion. This special feeling will hopefully be paralleled in radio if some sort of challenging DX activity can continue.

The drift here should be obvious and give rise to the following question: what will happen to DXing—not strictly as long distance talking but as a competitive sport? The answer: significant and complex changes will occur with upcoming developments but, as always, change is never total or complete, and I see the possibility of a mix of differing types of radio activity. We shall now develop some thoughts on this.

DXing beyond the year 2000 will change at some point as has been suggested in these pages. The technologically developed countries will participate together as progress is made. Amateur communication between these countries will be easy due to the synchronous action of operating coding. Seeking and initiating a contact, a very important part of DXing as we know it, might well be phased out in the automated QSO of the future. Using computers, a call for a particular country may be locked in to a computerized station from that country, and then the QSO can begin. Computers will locate a station in the desired country without assistance from the operator. This is a far cry from tuning and hunting the bands or calling CQs in the blind. The other computerized functions will make the QSO itself easy which will also be quite different than conducting a contact under bad QRM, QRN, and QSB conditions. One sad aside might be the phasing out of CW if it becomes truly unnecessary. It will be mourned of course, but ultimately our feelings toward it will be similar to the way we now regard the wire telegraph.

As is to be expected, the whole world will simply not be able to develop and support this technology simultaneously. There will therefore, be many countries using "primitive" equipment with which a QSO actually takes place much as we know it today. As a result, there might well be formal or informal frequency groupings, or both, that allow communication between modern and older stations. The sport of DXing could thus represent a combination of new technology for those countries with current communication systems and of old technology for those with earlier equipment. There will obviously be some variable "middle ground" operating technique as well while different generations of equipment are phased in and out. Thus, there will have to be a range of operating possibilities for several decades until the transformation is complete. There will be a limit, however, and this will go on for only so long. Then, the entire world will have changed. Finally, in the very distant future, communication will be altered so drastically that reliable speculation here as to its exact nature is not possible.

It is thus obvious that DXing will evolve and change. And yet, as time marches on, the spirit that marks the DXer of today will be evident in the future. The quest will never cease to be exciting, especially as "DXing" expands from earth to our own space and planetary stations. And, of course, we will always be in search of

Considered one of the most technologically advanced, single-sideband facilities in the world, Comm Central in Cedar Rapids (with a remote California site) utilizes groupings of computer-controlled Collins HF-80 equipment and 20 acres of antennas. Included in its multitude of world-wide communications functions is experimental high-frequency research.

the signal representing other life from deep in the cosmos.

All in all, DXing—both today and in the future—represents the result of tremendous amount of individual commitment and energy. Behind every DXing achievement, from personal to global from the efforts at one's own station to the efforts of activating a rare spot, is a considerable amount of enthusiasm and positive, creative thinking *and* doing. And that, my friends, is what DX POWER is all about!

Index